COMPLETE GUIDE TO
Home & Auto Burglar Alarms

By Doug Kirkpatrick

Complete Guide to Home & Auto Burglar Alarms

By Doug Kirkpatrick

Baker Publishing
Post Office Box 8322
Van Nuys, CA 91343

Copyright © 1986
By Doug Kirkpatrick
All rights reserved

No part of this book may be reproduced or transmitted in any form or by any means, electronic or mechanical, including photocopying, recording, or by any information storage and retrieval system without written permission from the publisher, except for the inclusion or brief quotation in a review.

Library of Congress Cataloging-in-Publication Data

Kirkpatrick, Doug
    Complete guide to home & auto burglar alarms.

    Includes index.
    1. Burglar-alarms. 2. Electronic security systems.
3. Automobiles--Anti-theft devices. I. Title.
II. Title: Complete guide to home and auto burglar alarms.
TH9739.K572 1987   643'.16   86-71338
ISBN 0-913193-02-X (soft)

FIRST EDITION
10 9 8 7 6 5 4 3 2 1

CONTENTS

Introduction

Why You Need This Book 7
Home Burglaries & Vehicle Theft
What You Can Do To Protect Your Home & Auto

Part One – Vehicle Alarm Systems

1 – It Really Can Happen To You! 11
Why Vehicles Are Stolen
Who Are These Car Thieves?
How Vehicles Are Stolen
What You Can Do

2 – Protecting Your Car, Truck & RV 17
The Purpose Of An Alarm System
How A Vehicle Alarm System Works
Alarm Components And What They Do
Using And Operating Your Alarm System
Summary

3 – How To Install Your Own System 37
Getting Started
What You Will Need
Where To Find Do-It-Yourself Supplies
How To Choose Alarm Components
Installing The Components
Wiring The Alarm System
Testing The System

4 – Other Vehicle Security Devices 49
Vehicle Immobilizers
Securing Hoods & Trunks
Window Glass Etching
Protecting Radios & Tape Decks
Protecting Wheels & Tires

Locking Gas Caps
Alarm Warning Window Stickers
Other Precautions & Considerations

5 – Professionally Installed Alarms 59
Evaluating Your Needs
How To Find A Professional Installer
How To Choose An Installer
Questions To Ask

Part Two – Home Security Systems

6 – Protecting Your Home & Family 67
Basic Protection Systems
Special Protection Systems
How An Alarm System Works
Basic Components
How Alarm Signals Are Transmitted
How Emergency Help Is Summoned

7 – Alarm Components & How They Work .75
The Master Control Panel
How The Control Panel Receives Messages
Circuit Characteristics Of Control Panels
Other Features Found In Control Panels
Battery Backup
Access Panels
Switches & Sensors For Perimeter Protection
Sensors & Detectors For Area Protection
Comparing Area Sensors
Special Protection Switches & Sensors
Tamper Switches
Alarms & Alarm Transmitting Devices

8 – Planning Your Security System 109
Determining Your Security Needs
Levels Of Protection
Summary Of Security Needs & Response
Planning An Alarm System For Your Home
Perimeter Protection – Sample Floor Plan
Area Protection – Sample Floor Plan

9 – Do-It-Yourself Installation 123
    Permits & Licenses
    Shopping For Alarm Components
    How To Choose Alarm Components
    How To Purchase Alarm Components
    Installing The Alarm Equipment
    Tools
    Supplies
    Installing:
        The Control Panel
        The Automatic Telephone Dialer
        The Digital Communicator
        Access Panels
        Local Alarms
    Installing Perimeter Switches & Sensors
    Installing Area Protection Sensors & Detectors
    Some Don'ts When Installing Area Sensors
    Installing Panic Switches
    Installing Heat & Smoke Detectors
    How To Wire Your Alarm System
    Sample Wiring Diagram
    Testing Your Alarm System
    Periodic Testing & Maintenance
    False Alarms
    Alarm System Warning Signs

10   Professional Alarm Systems 166
    How To Find An Alarm Company
    How To Choose An Alarm Company
    Underwriters Laboratories & The Security Industry
    Questions To Ask
    How To Purchase An Alarm System
    Alarm System Costs
    Paying For The Alarm System
    Final Walk-Through

Summary 173

Terms & Definitions 175

Index 185

Acknowledgments

To mention all of the people that made this book possible would take more space than is available. This list would also include various government departments, libraries, alarm system wholesalers and retailers. Special thanks to Carol Fong for her help and for the many suggestions she made.

Why You Need This Book

Most of us can remember when there was no need to lock our homes or worry about our cars being broken into. But now, even residents of small cities and towns are finding it necessary to take security precautions that were unheard of only a few years ago. Sadly, home burglaries and auto thefts are common place in today's society.

FBI statistics show that 90% of the reported crimes involve property. Of those crimes the most popular are burglary, larceny and auto theft. Last year alone 3 million burglaries were committed while 1 million vehicles were stolen. (These numbers don't include crimes that went unreported!)

These statistics are staggering, but they really don't mean anything until *your* house is burglarized or *your* car is stolen. Having something like this happen to you is upsetting to say the least. It is expensive and time consuming. If you've ever gone through the experience, you know the frustration of settling a claim with your insurance company. Trying to get reimbursed the full amount of your loss is next to impossible. In today's fast paced society, most of us don't have the time nor temperament to deal with these problems.

This book shows you what you need to know about buying and installing both home and auto burglar alarm systems. Whether you decide to install an alarm system yourself or have it done professionally, you'll see how to get the most for your money and learn how to get the right system for your needs. You'll also see how to lower your insurance premiums, while gaining the peace of mind of knowing your family is safe and your possessions are secure.

PART I

Vehicle Alarm Systems

Dedication – Part I

To Don, Al, Stan, Brenda, Dave and everyone else I know, who has had their car broken into or stolen at least once.

Chapter 1

It Really Can Happen To You!
Auto Theft

Do you know where your car or truck is right this moment? Are you sure that it will be exactly where you left it? Will your stereo tape deck still be there when you return? Don't get up yet, just read on.

Last year, vehicle thefts and break-ins accounted for almost $1.6 *billion* in losses. Those were only the ones reported to police. Who knows how many thefts and break-ins actually occurred! Needless to say, stolen cars, trucks and automotive accessories are big business.

If you own a BMW, Corvette, Porsche, or Mercedes you are probably aware that your car is on the hit list of professional theft rings. But, you probably don't realize that *any* car or truck is vulnerable to being broken into and stolen. Even Datsuns, Toyotas, Hondas, Chevys, and VWs are targets.

WHY VEHICLES ARE STOLEN

Vehicles are stolen for a variety of reasons, here is why:

Joy Riding - Joy riders will steal a car or truck just for the sport of it. Whether they are looking for excitement or do it on a dare, it doesn't make any difference to the owner. A thief takes property without concern for its return or well being. Hopefully, the police will find it undamaged.

When a vehicle is stolen for joy riding, the make or model doesn't make any difference to the thief. Any car will do as long as it can be easily started. The joy rider looks for any opportunity where the owner has forgotten to take the key

Vehicle Alarm Systems

or has left the car running while doing a errand. In this case, even the best alarm system won't help. The owner simply has to remember to lock the car and take the key every time!

Because of ignition and steering column locks, joy riding is not as popular as it once was.

Used To Commit Another Crime - A criminal with any intelligence wants a getaway car that cannot be traced back to him, in case someone happens to get a description or the license number. Again, the thief is not particular about make or model as long as the stolen vehicle serves his transportation needs. He looks for an easy target but won't hesitate to break the steering column lock or hot wire the ignition.

For Resale - That's right, for resale. Thieves want your car so they can sell it, especially if it is a late model or classic. Professional rings operate in a number of ways. One method is is to change the Vehicle Identification Number or "VIN". What happens is that the stolen vehicle assumes the identity of a similar junked or wrecked car. Paint and other cosmetic changes are made so the car cannot be recognized. The "wreck", which is actually stolen, is then sold to a legitimate and unsuspecting buyer.

Another technique is to sell the stolen vehicle in another state using forged ownership papers. This technique works because some states and foreign countries are lax in checking documents allowing a thief to disappear before anyone realizes what has happened. Many times a professional fills orders for specific makes and models. When a request is made, the pro finds a car and then sells it to someone who forges identification papers or sells the vehicle part by part. A professional thief sometimes steals a car, strips it, and then buys it back from the insurance company to reassemble it.

For Parts And Accessories - You may have heard the expression, "The whole is greater than the sum of its parts". But with some sports cars, late models and classics, just the opposite is true. Parts and accessories taken from stolen vehicles and sold piece by piece bring a much larger return. Also, parts that cannot be traced lessen the risk of being caught.

Auto Theft

A skeletal shell is all that remains after a car has been through one of these "chop shops". The body and engine parts are then fenced through "legitimate" body shops, auto repair garages and at swap meets.

Sound Systems Specialists - Some thieves specialize in stealing only radios, amplifiers, tape decks and speakers. A specialist can break-in and take your entire sound system in minutes. He literally breaks the radio out of the dash doing hundreds of dollars worth of damage to the vehicle's interior.

Valuables Left In The Vehicle - Sometimes what is in a vehicle is more valuable than the vehicle itself. For example, a truck or van filled with tools, merchandise or other items will attract the attention of a thief. The vehicle is only needed to transport the goods to another location.

WHO ARE THESE CAR THIEVES?

Car thieves are primarily young. Arrest records show that 65% of all persons arrested for motor vehicle thefts were under the age of 21, and almost half of those were under 18. Many are joy riders, but an increasing number are committing these crimes for money. Motives of thrill and adventure are being replaced with greed and illegal gains.

Another factor adding to the problem is the organized rings of professionals. The difference is that the amateur is an opportunist while the professional plans his jobs very carefully.

With the price of cars and parts, motor vehicle theft is becoming big business!

HOW VEHICLES ARE STOLEN

Before a thief can steal possessions, accessories or the car itself, he has to get inside. But, most thieves don't have the time nor the skill to pick a door lock. It is much easier using burglar tools or brute force. (It is even easier if the owner has forgotten to close the windows and lock the doors!)

Vehicle Alarm Systems

Professional thieves use a tool called a "Slim Jim". This tool slides down into the door panel and attaches itself to the door locking mechanism. With practice, the thief can get inside in a matter of seconds.

By contrast, amateurs will use a screw driver, crow bar or rock to bend, break or smash their way in. Sometimes the cost to repair this damage is more than replacing what was stolen.

Once inside, the thief has a number of ways to get a vehicle started. Here is how:

Hot Wiring - Before car manufacturers were required to install locking steering columns, it was very easy to start a car by jumping the ignition wires under the dash. The good news is that hot wiring is not as popular as it once was. The bad news is that thieves today have better ways to start a car very quickly.

Ignition and Steering Column Lock Removal - Even if your vehicle is equipped with an ignition and steering column lock, it is still vulnerable to a thief. A thief simply extracts the ignition lock or breaks it out of the column. A pro can get inside, remove the lock and then start the car in less than a minute.

Towing - Some professional rings won't even bother trying to start a car. They disguise themselves as legitimate tow truck operators and haul their "loot" away. They have little trouble taking a car without attracting attention.

Master Keys - The easiest method of stealing a vehicle is to have its keys. With a key cutter, key blanks and the key identification number, a thief can make his own key. Unfortunately, these items are not difficult to obtain.

As you can see it is not hard to break into a vehicle and steal it. My purpose is not to scare you, but to make you aware of what you and your vehicle are up against. As you read on you'll see how to fight back.

WHAT YOU CAN DO

You can usually stop an amateur or at least slow him down, but it is almost impossible to stop a skilled and determined professional. If he wants your car he can probably get it.

So why try to protect your car if it won't do any good? The answer is that you are trying to deter a thief. By installing an alarm system and by taking other precautions, you are sending a message that *your* vehicle is going to be very difficult and time consuming to break into. And if a thief does get inside, he will attract a lot of attention while risking arrest. Hopefully, the thief goes elsewhere leaving your car alone.

The next chapter shows you how the various alarm systems work and what is available to protect your car, truck or RV.

Chapter 2

Protecting Your Car, Truck & RV

Because there are so many different alarm systems for cars and trucks, it's mind boggling choosing the right one. Manufacturers are producing numerous alarm systems with endless equipment combinations. So how do you make the right decision when it comes to your vehicle?

First you have to decide how much protection you need and how much you can afford to spend. Then you have to find out what is available and whether you will install your own system or hire someone to do it for you.

In this chapter you will see the various alarm systems and find out how they work. But before getting there, let's take a closer look at what an alarm system actually does.

THE PURPOSE OF AN ALARM SYSTEM

A vehicle alarm system is designed to do one or more of the following:

1) It scares away would-be thieves by sounding a warning siren, bell or horn.

2) It attracts attention to the break-in while alerting someone to call the police.

3) It disables the ignition, fuel system or starter, preventing the vehicle from being driven away by a thief.

Vehicle Alarm Systems

4) It sends a message through a paging transmitter and receiver warning you that someone is tampering with your car.

Simple alarm systems only sound a warning alarm while a more sophisticated system sounds the alarm, disables the vehicle and sends you a message, through a pocket pager, that someone is tampering with your car. The system you choose to install will depend on factors such as what kind of vehicle you are trying to protect, where you park it and how much you can afford to spend.

Remember, you want to make it difficult and time consuming for anyone wanting to steal your car. Your job is to deter the thief so he will leave your car alone. Make his job as hard as possible and he will go somewhere else.

HOW A VEHICLE ALARM SYSTEM WORKS

First, the alarm system must detect a disturbance from a would-be thief. This disturbance can be anything from opening a protected door to shattering a window. Once a switch or sensor is activated, a message is sent to the control panel or unit. The panel then responds by sounding a bell or siren.

In addition, the control unit can also transmit an alarm message to you via a paging system. The diagram on the following page shows you what this process looks like.

HOW AN ALARM SYSTEM WORKS

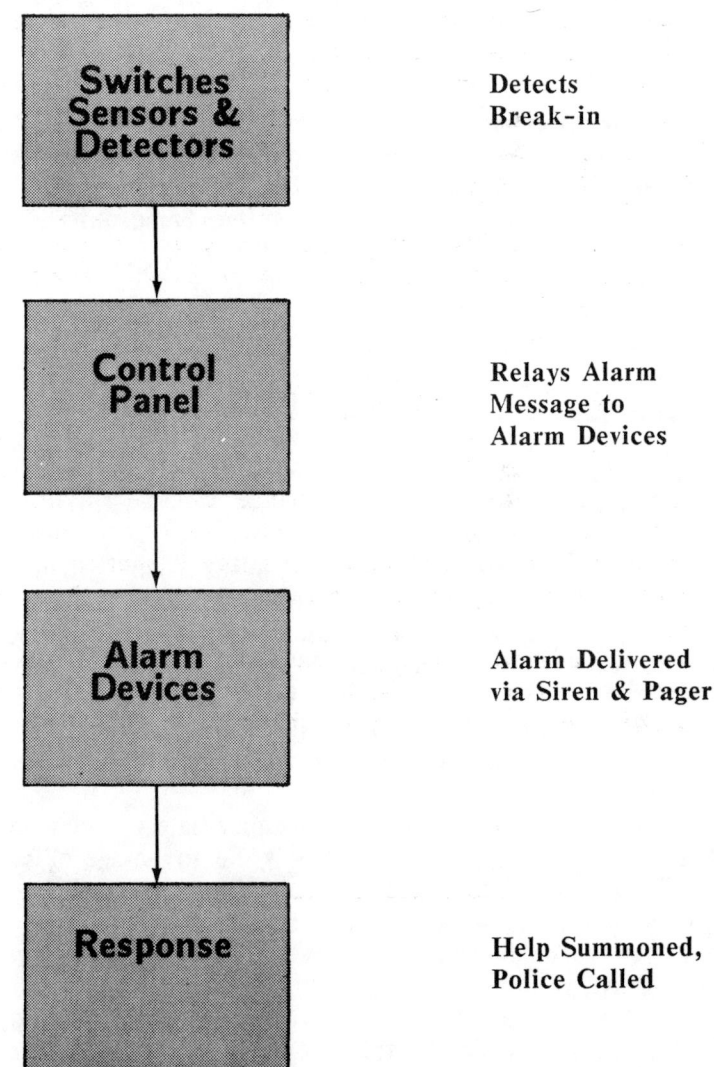

Vehicle Alarm Systems

Now, let's describe this process in more detail.

Switches and Sensors - Their purpose is to detect a thief tampering with your car. Some are mechanical such as a pin switch which detects a door being opened. Voltage drop sensors monitor electrical currents. Motion detectors sense when a vehicle is being jacked up or towed. The job of a switch or sensor is to activate the alarm system.

The Control Panel or Unit - The control panel is the nerve center of the alarm system. It receives messages from switches and sensors and then sends the alarm message through the siren, bell and/or paging transmitter. Some units also have circuits and relays for disabling the ignition system, electrical fuel pump or starter motor.

Alarm Devices - Vehicle alarm systems use either electronic sirens, mechanical sirens, bells or horns. Their purpose is to attract attention, to scare away a thief and to have someone call the police. In addition to sounding a siren or bell, some alarms activate the vehicle's head lights giving a visual warning as well.

Paging Systems - Paging systems are silent alarm devices. If a thief trips a sensor or switch, a coded message is transmitted to a receiver that can be carried in your pocket or purse. When an alarm message is sent, the receiver beeps and flashes a warning that your car is being tampered with. Depending upon the transmitter's antenna and the surrounding terrain, you can be up to eight miles away and still receive the alarm message.

Paging systems notify you immediately when someone tries to break into your car. This is the advantage of having a pager in addition to a siren or bell. With only a siren or bell, you would have to depend on the siren or bell to scare the thief away and have someone else call the police. (How many times have you heard and ignored a blaring auto alarm?)

If you install a paging device and it goes off, don't try to stop or apprehend the thief yourself. It could be potentially dangerous. It is better to call the police and let them handle the situation.

Protecting Your Car, Truck & RV

Response - Hopefully, the siren, bell or horn will scare off the thief. If not, hopefully someone will see what is happening and call the police. But you really cannot depend on someone else. This is why you might want to consider a paging system making sure the police will be notified.

ALARM COMPONENTS AND WHAT THEY DO

In the previous section you saw the basics of vehicle alarm systems and how they summon help. It is now time to take a closer look at the components that make up a system.

The Control Panel

The control panel or unit is the center of an alarm system. It is made up of solid state components, circuit boards, timers and relays. All of the switches, sensors, detectors, sirens, horns, bells and paging units are connected to it. When this unit receives an alarm message from a sensor, it sounds the warning devices while transmitting to the paging receiver.

Lock Up Or Latching Relays - These relays maintain an alarm once the system is tripped. If a thief enters through a door, a relay prevents the system from shutting down when the door is closed.

Delay Circuits - Delays are needed if your alarm system is turned on and off from inside your vehicle with either a key pad or hidden switch. The delay circuit gives you time to activate your alarm system and then exit the car before the system will respond to a break-in. Upon returning, the delay gives you time to deactivate the system before the warning alarm is sounded. Delay circuits are used for switches on the driver's door. All other switches and sensors are connected to an instant alarm circuit in the control panel.

Instant Circuits - In contrast to the delay circuits, instant circuits activate the alarm system and sound a siren or horn immediately. There is no delay from the time a switch or sensor detects a disturbance to the time the alarm goes off.

CONTROL PANEL

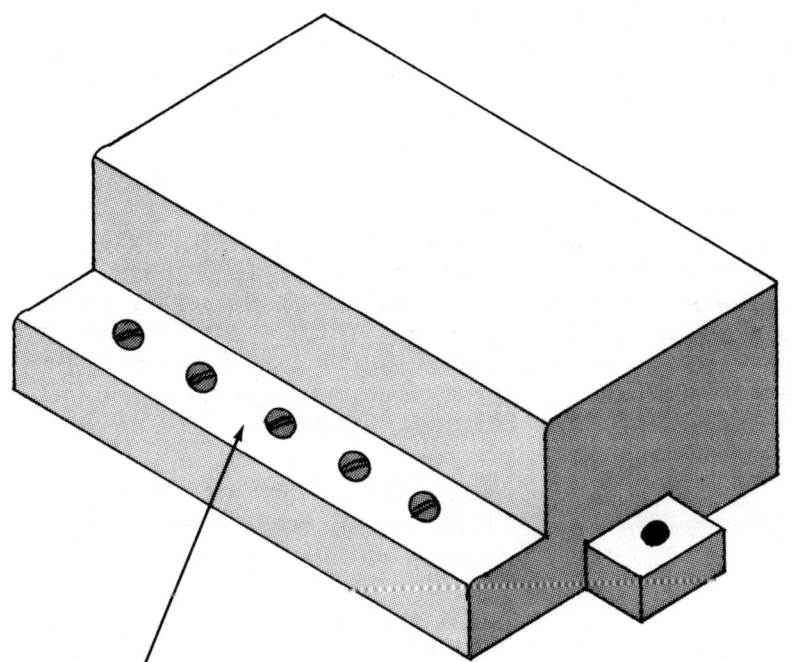

Connectors for:

    Switches, Sensors, Detectors
    Alarm Bell or Siren
    Paging Antenna
    Arming/Disarming Switch
    Ignition Cutoff
    Power Lead from Battery

Additional Circuits - Special circuits are for accessories such as stereo tape decks, citizen band radios and mobile telephones. In addition, they can also be used to protect accessories on the outside of your vehicle. These include luggage racks, ski racks and fog lights. If one of these items is disturbed by a thief, an alarm is sounded.

Reset Or Cancel Switches - Once the alarm system is activated, a reset switch cancels an alarm and prepares the system to detect another disturbance. These switches are especially useful if you accidentally trip the system and need to shut off the alarm siren or horn.

Automatic Reset - When your alarm system is activated, it will sound indefinitely unless equipped with an automatic reset. Once the switch or sensor is returned to a non-alarm position or condition, the automatic reset shuts off the alarm after a predetermined time and then prepares it to detect another attempted break-in. Most cities require this feature to prevent alarm sirens and bells from sounding until your battery goes dead.

Override Or Valet Switches - A override switch allows you to temporarily shut down the alarm system. This is important when you let others drive your vehicle.

Powering The System - Almost all alarm system control boxes are powered by the 12 volt battery from the vehicle. The more sophisticated and expensive alarms also have a backup battery just in case the vehicle's primary battery goes dead or if it is disabled by a thief. It should be mounted in an inaccessible area so it cannot be reached.

Ignition Cutoff

Most sophisticated control units have this feature. It is important to look for when buying an alarm system because it is one of the best ways to defeat a professional car thief. Even if he has a master key or pulls out the ignition lock, the car cannot be started.

Vehicle Alarm Systems

Some cutoff features allow a car to be started but then disable the ignition after a short time. The idea is to take the thief by surprise. When he cannot restart your car, he is forced to abandon it in the same area where it was taken.

There are drawbacks to interrupting the ignition system. The ignition lockout was a relatively simple device used on cars built prior to 1973. But now most cars have electronic ignitions and putting an ignition lockout device on your car might technically be illegal. This is because it could change the emission controls. In addition, if the alarm system malfunctions, your car could cutout in traffic creating an inconvenient if not dangerous situation.

Starter Lockout - As an alternative to disabling the ignition system, some alarm systems will interrupt the circuit between the battery and the starter. This prevents the car from starting, but doesn't create a dangerous situation if the alarm system malfunctions while driving.

Electronic Fuel Pump Lockout - Here is another option for disabling your vehicle. Instead of interrupting the ignition system or starter motor, interrupting the electrical fuel pump does basically the same thing. You won't have emission control problems, but will have stalling problem if the alarm system malfunctions while driving.

Switches, Sensors & Detectors

Vehicle alarm systems detect tampering and break-ins by monitoring the perimeter of the car with pin switches, motion detectors, sound discriminators and voltage drop sensors. Other sensors and detectors will only trigger an alarm after the thief is inside the car. This is referred to as space or area protection. You might want to install either perimeter or space protection sensors, or a combination of both.

There are quite a variety of devices for detecting thieves tampering or breaking into your vehicle. The better systems have more than one type of device to detect different threats. Here are the various switches, sensors and detectors you will see:

Protecting Your Car, Truck & RV

Pin Switches - You have probably seen pin switches in the door frames of your car. They are used to turn on and off the dome light in the passenger compartment. When connected to a protection circuit in the control unit, they initiate an alarm when a door, hood or trunk is opened. Pin switches are inexpensive and not too difficult to install.

Motion and Shock Detectors - If your vehicle is jacked up, bumped, hit, towed, or moved, these detectors set off the alarm. (For those of you familiar with home security systems, the term "motion detector" has a different meaning as it is used for vehicle security systems.) Motion and shock sensors allow you to adjust the sensitivity so as to minimize false alarms.
Mechanical motion or shock sensors use weighted contact points which are kept apart by a spring. When the vehicle is bumped, jacked up or moved, the weight moves the contacts together completing the alarm circuit.
The more sophisticated motion sensors are electrical or use mercury to complete the circuit.

Sound Discriminators - Sound discriminators have been used in home security systems to listen for burglars breaking in. Recently, they have been adapted to auto security systems.
A discriminator listens for high pitched sounds such as glass breaking or when metal is pried. They react when a thief breaks a window or tries to pry the door, trunk, or deck lid.

Voltage Drop Sensors - Drop sensors connect directly into the vehicle's electrical wiring system. They detect very small voltage drains such as a door opening and turning on the dome light.
Many professional installers don't recommend voltage drop sensors. Instead, they install pin switches which they claim work better, are more precise and less subject to false alarms.
These sensors are also not practical for recreation vehicles. Because of the various accessories which start automatically, they can initiate a false alarms.

PIN SWITCH

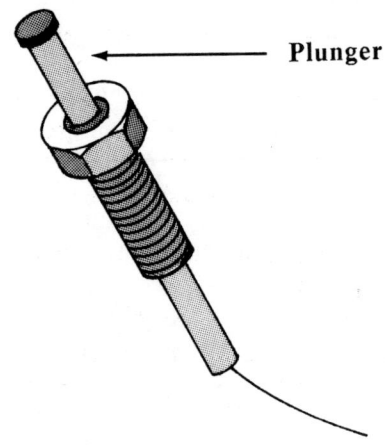

MOTION DETECTOR

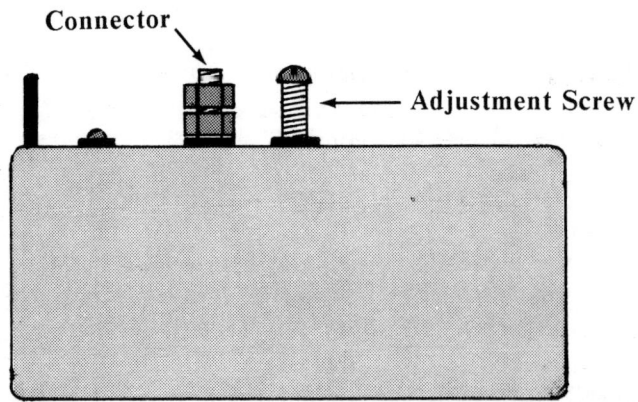

Protecting Your Car, Truck & RV

Pressure Sensitive Mat Switches - Mat switches are installed in the seats, under the standard floor mats or below the carpeting. An alarm is triggered when someone sits on the seat or steps on the car floor. Pressure sensitive mats are installed in convertibles and other open cockpit vehicles where other sensors and detectors would be impractical.

Ultrasonic Sensors - Ultrasonic sensors fill the passenger compartment with inaudible sound waves. When a thief violates this area, the sensor recognizes a disturbance in the sound wave patterns and then initiates the alarm.

Panic Switches - With a panic switch you can trigger the alarm system yourself. This can be especially important if a thief or mugger approaches when you are in or just about to get into your car. The panic switch can be hidden under the seat or can be on a key ring transmitter.

Alarms

Alarm devices deter thieves by attracting attention. Noise from the siren, bell or horn will usually scare the thief away before he can start your car or take any of the accessories.

Sirens - Two types of sirens are used in vehicle alarm systems: electronic and mechanical. Electronic sirens consist of an electronic circuit to convert electrical impulses into a warning sound which is delivered through a speaker. Mechanical sirens have a fan that pushes air through the siren creating the sound.
Electronic sirens tend to be more reliable than mechanical because they have no moving parts. In addition, electronic sirens can vary the pitch and frequency, making their sound much more noticeable than a continuous tone from a mechanical siren or bell. This change in the sound is referred to as a "yelp", "rise and fall" or "high/low".
Sirens are rated in decibels (dbs), which is the measure for sound levels. Simply put, the higher the dbs, the louder the noise. The sound level will double for every 3 dbs. For

ALARMS

Siren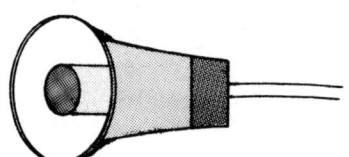

Bell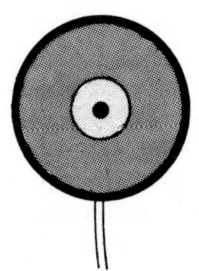

Air Horn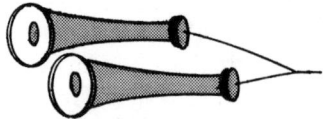

example, 93 dbs is twice as loud as 90 dbs and four times as loud as 87 dbs.

Here are examples of every day noises as rated in decibels:

Jet Aircraft	**140 dbs**
Car Horn	**89 dbs**
Conversation	**60 dbs**

Many sirens for vehicle alarms are in the 90 to 110 decibel range.

The frequency of the sound is just as important as the loudness. Low frequency sounds carry better but high frequency sounds are more distinct. This is why a high/low siren is more noticeable than a continuous tone.

Sirens are usually installed under the hood and in protected areas so that a thief can't reach and disable them. Some installers will also put a mini siren inside the passenger compartment to scare a thief that gets inside before the outside alarm sounds. The inside siren is to scare the thief while the outside alarm attracts attention.

Other alarm installers will use both an electronic and mechanical siren. When the alarm goes off there are two separate sounds to confuse the thief and that makes it twice as hard to find and disable the alarm.

Bells - Bells are not used as often as sirens in alarm installations. They have moving parts and are more likely to fail than an electronic siren.

Air Horns - These alarm devices are becoming more and more popular because they are extremely loud. Many times they are installed in addition to a siren. The idea is to create two distinctive warning sounds which will certainly attract attention.

Voice Synthesizers - One of the latest ideas of scaring off a car thief is to install a system which simulates a human voice. It shouts "BURGLAR! BURGLAR!" or other warnings when someone tries to break into your car.

Vehicle Alarm Systems

Car Horn - Inexpensive alarm systems use the car's horn as the only warning device. The more sophisticated alarms will use a siren as the primary warning sound and the vehicle's horn as a secondary sound.

Headlights - Here is another option to look for in alarm systems. Some alarms units connect to the headlights, turning them on and off if a thief tries to break-in. The pulsating headlights give a visual warning to compliment an audible warning from the siren, horn or bell.

Paging Systems

A paging system consists of a transmitter and a receiver. The transmitter can be a part of the control panel or an add-on unit. The receiver is a small battery powered unit that you carry in your pocket or purse. When the alarm system is triggered, the transmitter sends a coded signal to the receiver.

The radio signal referred to above is transmitted via your standard radio antenna with a range of up to 1/2 mile. By installing a citizen band radio antenna, you can receive an alarm message up to 8 miles. This distance will depend on the local terrain and atmospheric conditions which affect radio transmissions.

You have to be careful if you park near high-rise buildings or in underground garages, as these structures affect the reception of radio waves as well. It is possible to be within 100 yards and not receive a transmission. You also have to periodically check the batteries in the receiving unit to make sure they can receive the alarm message if one is sent.

With thousands of different combinations, it is unlikely that your transmitter and receiver would be operating on the same frequency and in the same area as another.

Remember, if you have a paging system and receive an alarm message, be extremely careful. Don't try to stop or apprehend the thief yourself. Who knows, he might be armed. It is best to call for the police than to confront the thief. (Unfortunately, the police are sometimes slow in arriving, especially if the call is not a life threatening situation. Don't make it one by trying to be a hero!)

PAGING SYSTEM

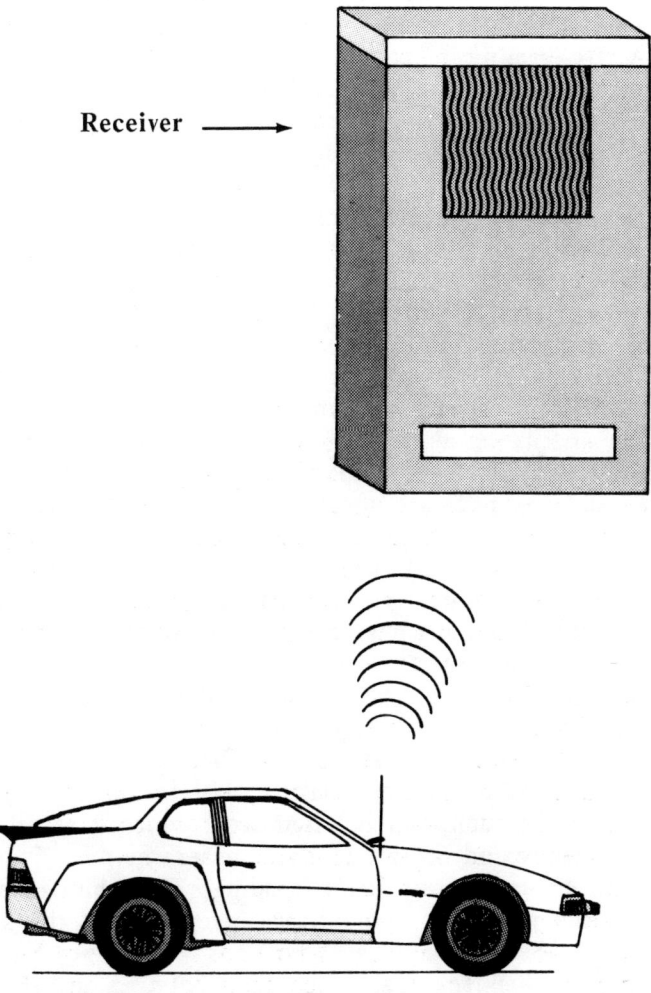

Vehicle Alarm Systems

USING AND OPERATING YOUR ALARM SYSTEM

Arming the System

Once you have parked your car, you now need some method of arming the alarm system. Vehicle alarms have two basic ways of doing this; either *actively* or *passively*.

Active arming means that you must manually activate your alarm system making it ready to detect a break-in or attempted theft. An actively armed system can use a key lock switch, hidden toggle switch, digital key pad or transmitter for arming.

Key lock switches are mounted on the outside of the vehicle, usually on the left front fender. They can be hidden from sight in a wheel well, or they can be in full view. In full view, they act as a deterrent to some car thieves, but only warn the professional thief to be more cautious.

Hidden toggle switches mounted inside the passenger compartment are easier to install and are more secure. A thief will have to trip the alarm before getting a chance to find the switch and override the system.

Digital key pads are also used for arming and disarming. Mounted inside the passenger compartment, they are more convenient to use and don't require carrying an extra key. All you do is enter your personal code to turn on or off the system. Digital key pads have alarm status indicator lights that tell if your system is armed or disarmed.

Key pads that can be mounted on the dash or console are a visual deterrent to thieves as well.

One of the latest methods of arming and disarming a system is with a **key ring transmitter**. The transmitter puts out a coded radio frequency or infrared beam which is received by a small antenna or infrared light sensor connected to the control panel. Instead of using a key pad or hidden toggle switch in the passenger compartment, you control the system from outside the vehicle.

If you consider an alarm system that uses one of these transmitters, make sure that the transmitter cannot activate the alarm system while the car is in motion. This is especially important if your system uses an ignition interrupter. You

KEY LOCK SWITCH

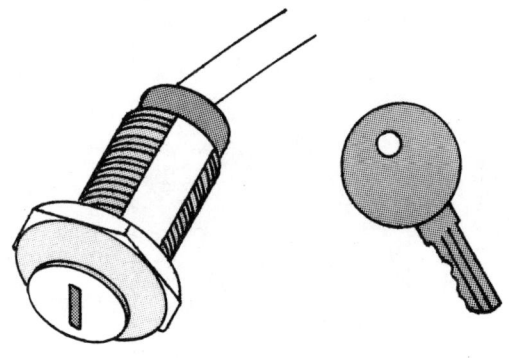

TOGGLE SWITCH

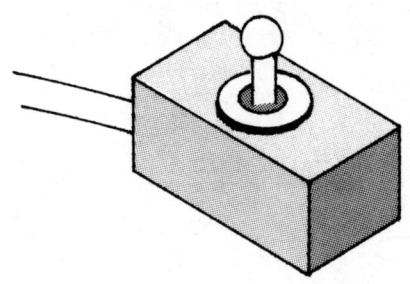

DIGITAL KEY PAD

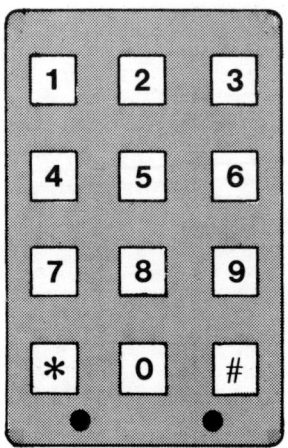

KEY RING TRANSMITTER

Protecting Your Car, Truck & RV

could accidentally shut off the motor making the car very hard to steer or stop.

The transmitter is small enough to be carried on a key ring and allows you to disarm the system from outside the vehicle. This is important because a delay circuit is not needed. (Remember, if you arm and disarm your system from inside the car, you need a delay so you won't set off the system when you get in and out of your car. Any delay might give a thief enough time to find and disable the alarm system before it sounds its warning.)

In addition, a transmitter can also be used as a remote panic switch and can lock and unlock your car if it is equipped with power door locks. This option is helpful especially when your arms are full of packages.

While these transmitter systems are sophisticated and make your alarm system harder to compromise, they do have their disadvantages. You have to carry the transmitter and periodically check the batteries so you can always disarm your system.

Passive arming - The problem with a actively armed system is that you have to remember to set it. The solution for car owners who might forget is a passively armed system. With a passive system, once the ignition key is removed, the system sets itself. The system delays before arming, allowing you time to get out and close the doors. The better alarm systems "chirp" after arming, giving you an audible acknowledgement that the system is ready. This reassures you that the system will sound an alarm if needed. Other systems show that the alarm is activated by turning on a small light, usually located on a key pad.

Because you can't forget to set the alarm, passively armed systems usually qualify for the highest insurance premium discounts available.

Disarming the System

When you return to your vehicle, you have to have some way of disarming the system so as not to set off the alarm devices and to restore any ignition, fuel, or starter cutoff circuits.

If you actively arm your system with a key lock switch, hidden toggle switch, digital key pad or transmitter, you will

Vehicle Alarm Systems

probably disarm it the same way. If you have a passively armed system, you will also need one of these devices to disarm it.

SUMMARY

In this chapter you have seen the basics of vehicle security systems. I hope you now have a better understanding of how a security system can protect your car and possessions. If you are handy and enjoy working on your car, maybe you should consider installing a system yourself. The next chapter shows you how.

Chapter 3

How To Install Your Own System

In the previous chapters you have seen the basics of vehicle alarm systems. And you probably have some idea of what protections you need to install. This chapter shows you how to install your own alarm system and save money.

With some mechanical ability and electrical skill, it really isn't too hard to do your own installation. Most of the alarm systems for the do-it-yourselfer come with easy to follow installation instructions. But, if you don't know which end of a screw driver to use, you will be better off finding someone else do the installation for you.

GETTING STARTED

What To Do First - Again, start by evaluating your security needs. Think about where you park your car while at home and at work. Also remember that you have to leave it unattended while shopping, at restaurants and hundreds of other places. A professional car thief can steal it in under a minute, so don't assume it will be safe anywhere, even if you leave it unattended only a short time.

The make and model of your vehicle also makes a difference. If you own a car or truck which is a high theft item, you should have at least an ignition, fuel or starter cutoff and an alarm sounding device. You should protect every opening with a switch or sensor. This would include the doors, truck, hood and deck lid. You might want to install a paging system and backup battery as well. Remember, inexpensive

Vehicle Alarm Systems

systems do not deter professional car thieves. All you are installing is a false sense of security.

When planning your system it is a good idea to use a combination of different switches, sensors and detectors that will detect a variety of attacks. Protect all openings such as the doors, hood, trunk and deck lids. You will also want to leave a way to by-pass the system just in case it malfunctions.

One of the most important things to consider is just how you want to arm and disarm your alarm. You can choose among hidden toggle switches, key lock switches, digital key pads and remote transmitters.

WHAT YOU WILL NEED

Here are the various components, tools and supplies needed for do-it-yourself alarm systems:

1) Control Panel or Unit

2) Switches, Sensors, Detectors

3) Alarm Devices - Sirens, Bells, Horns

4) Toggle Switch, Key Lock Switch or Key Pad

5) Supplies - Wire, Electrical Tape, Solder

6) Tools - Screw Drivers, an Electric Drill Gun, Drill Bits, Wire Cutters, Soldering Gun

Also, try to get a wiring diagram for your car. You'll find it a real time saver when tracing unfamiliar wires. Wiring diagrams are usually included in the vehicle's shop manual.

WHERE TO FIND DO-IT-YOURSELF SUPPLIES

Here are some of the many places you can find alarm system components and installation supplies:

1) Automotive Stores

2) Discount/Variety Stores

3) Do-It-Yourself Home and Automotive Security Stores

4) Electronic Supplies Stores

5) Specialty Mail Order Catalogs (See page 124 for list)

HOW TO CHOOSE ALARM COMPONENTS

Vehicle alarm systems designed for the do-it-yourselfer are tested and evaluated in the various consumer magazines. You can sometimes find a write-up on a specific system you are considering. Check back issues of magazines such as **Changing Times, Consumer Reports, Popular Science, Popular Mechanics** and **Mechanics Illustrated**, for auto alarm system evaluations.

When choosing alarm components, make sure that the control panel, switches, sensors, detectors, sirens, bells and paging units will all be compatible. For example, one manufacturer's sensors might not work with another's control unit. The point is to ask questions and find out about any potential installation problems before you buy.

Some manufacturers put together packages or kits that contain everything you need to install an auto alarm system. They have a control unit, switches, sensors, wire, siren, paging transmitter and receiver, and a key pad or key operated switch as well as complete installation and wiring instructions.

Other manufacturers use a component approach. What they do is package and sell all the various items you need separately. This allows you to design your system and install just the switches and sensors you want. This component approach also gives you the advantage of adding additional devices to your system at a later time.

Vehicle Alarm Systems

INSTALLING THE COMPONENTS

The important thing to remember when installing any of the alarm equipment is to mount and place it so it cannot be reached and compromised by a thief. You must prevent access to the engine compartment. This is where vital components like the sirens, horns, power supply and wiring are located.

If a thief suspects an alarm and still wants to break-in, he will usually try to disable the system by either cutting the power or by disabling the siren or horn. In addition, the thief might try to override any of the ignition, fuel or starter cutoffs.

Most professionally installed vehicle alarms use a "dead bolt" latching mechanism which locks and unlocks the hood from inside the passenger compartment. If the do-it-yourself system you buy does not come with a dead bolt for the hood, you can always purchase one separately and incorporate it into your alarm system.

It is very important to read and understand all of the instructions that come with your alarm system. Make sure you know which wire to connect to which terminal before starting the installation.

Installing The Control Panel - The first thing to do is to find a location for the control panel. You will want to install it in the passenger compartment, under the dash board and away from the heater or air conditioner. If space does not permit you to mount it under the dash, try mounting it in the trunk or under the seat.

You also will want to make sure that a thief cannot reach the control unit from outside the vehicle. Some manufacturers recommend installing it in the engine compartment. However, this is not a good idea because the temperatures under the hood can reach well over 300 degrees Fahrenheit. In addition to heat, the control unit would also be subject to dirt, grease, moisture, and vibration. The idea is to keep it away from conditions that can damage the internal electronics.

Installing the Switches, Sensors and Detectors - Different sensors require different installation techniques. It is

TYPICAL ALARM INSTALLATION

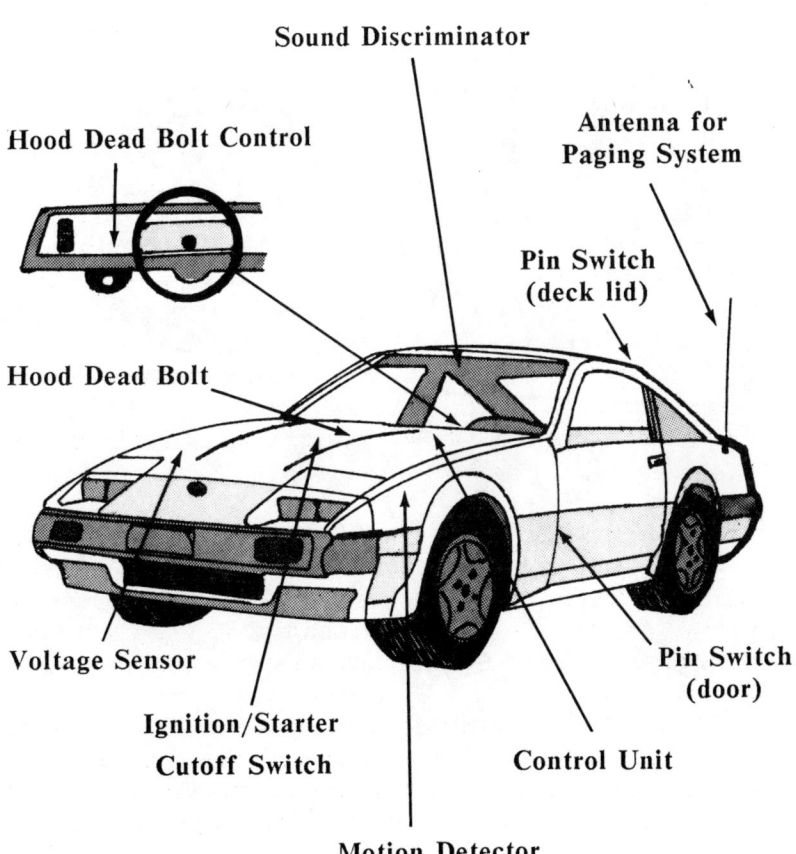

PIN SWITCH INSTALLATION

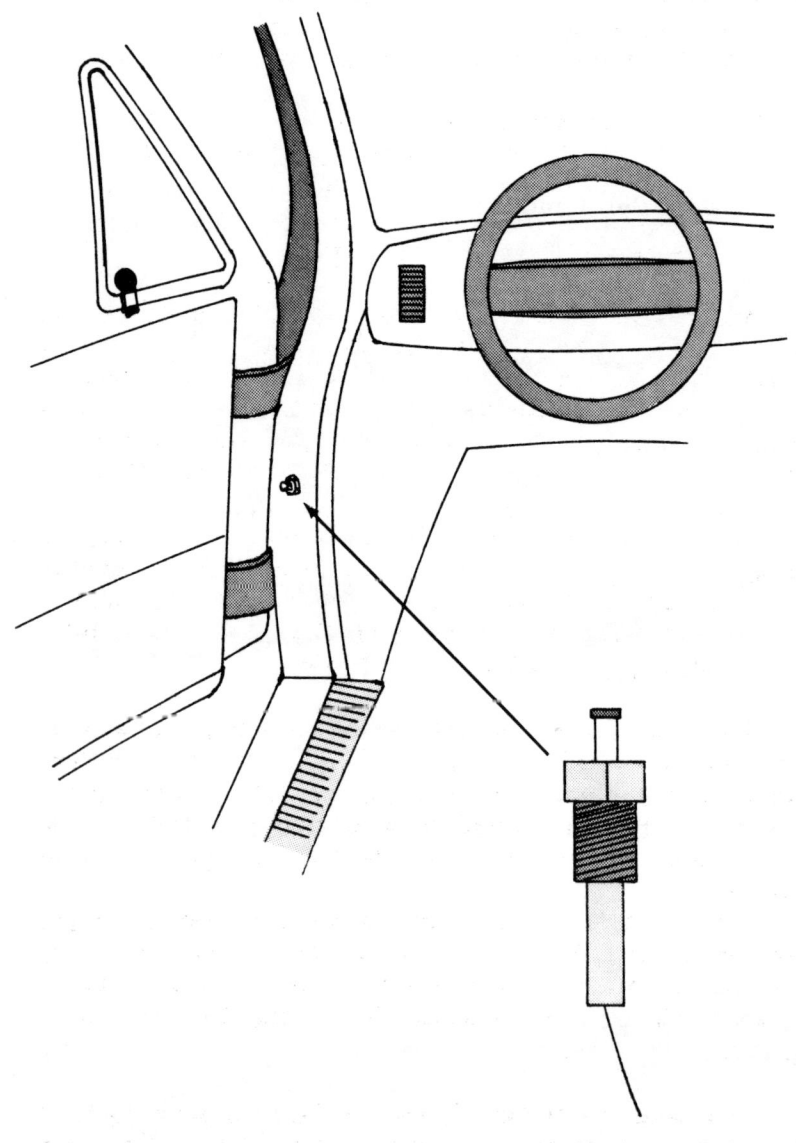

How to Install Your Own Alarm System

best to read all installation instructions carefully before trying to install any of the switches, sensors or detectors.

Some sensors and detectors such as sound detectors, voltage drop sensors and motion detectors are simple to install. Just decide where they should go and then secure them in place. Some sensors and detectors are attached with sheet metal screws while others can be secured with epoxy glue.

Caution - *You should always check before drilling any hole used to mount a control panel, switch, sensor or alarm device. Make sure you don't drill into a hidden wiring harness or other vital part of your vehicle.*

Pin switches take a little more effort to install. You will want to install one in every opening point, such as doors, hoods, deck lids, trunks and tail gate. Try to find self-tapping pin switches because they are the easiest to install. First, drill the holes for the pin switches in the door jambs and support brackets. Then insert the switches and gently screw them in. Make sure that each one grounds to the car frame.

After you have connected the wiring and tested the switch, it is a good idea to coat the switch with a lithium grease to prevent corrosion. Because pin switches are subject to wear and tear, you should also clean and check them periodically so they will work when needed.

Installing the Alarm Devices - Remember, you want to protect the alarm bell, siren or horn in a location where a thief cannot disable them. Install these devices in the engine compartment that is secured with a dead bolt lock on the hood. Mount them securely using the hardware provided with your kit.

Some installers also mount a siren in the passenger compartment. If a thief gets inside the vehicle through a door on a delay circuit, the inside alarm will take him by surprise. The noise will be very uncomfortable causing him to make a fast exit.

Installing a Paging System - Some paging systems incorporate the receiving unit with a control panel. You have

Vehicle Alarm Systems

to purchase a unit that allows you to connect the various switches, sensors, detectors and alarms that you want to use.

With a paging system you need to attach the receiver lead to an antenna. It is a good idea to mount the unit near the radio to minimize the antenna lead. Some pagers connect into the regular radio antenna lead. Some may require an additional antenna, especially if your antenna is encased in the windshield. For optimum performance manufacturers recommend installing a high gain CB antenna.

One problem with a paging system is if the antenna is broken off. This renders the pager useless. Another problem is connecting the pager to a power antenna which is normally retracted when the car is parked. It is possible to install a relay which raises the antenna if the alarm system is activated. But this is usually left to the professional installer. Instead, you might want to consider installing a separate antenna solely for the pager.

After installing the paging unit connect the antenna lead before you connect the power. Some units are damaged if they try to send a message without an antenna lead.

If you have a CB radio and want to use the same antenna, install a two way switch in the antenna lead to prevent transmissions between the CB and the pager.

Installing Key Pads And Key Switches - Depending upon your system, you need some way of activating and deactivating it. Even with passively armed systems, you have to have some way of turning it off.

The most convenient place to install a key pad is either on the dash or on the console between the front seats. Placing it in clear view reminds you to use your alarm system and also acts as a visual deterrent to thieves. It puts would-be thieves on notice that your car is protected.

Key lock switches are designed to be installed on the outside of the vehicle. Many installers don't like them because they could give a thief a chance to disarm your alarm before activating a switch or sensor. Others feel that the presence of the key lock acts as a deterrent. If the alarm system you buy only comes with a key lock switch, you have two choices: either you can mount the switch in a conspicuous location and hope it deters a thief, or you can hide its installation in a wheel well.

How to Install Your Own Alarm System

Use a cover on the key lock switch to prevent dirt and corrosion from jamming the lock.

Some alarm systems are activated and deactivated with a hidden toggle switch located under the dash board or seat. However, the alarm system must have a exit and entry delay, allowing you to turn on and off the system without setting off the alarm.

If a thief is smart enough to get into your car and find the switch, he could disable the system before the alarm sounds. One trick some installers use is to install more than one hidden switch. By wiring a number of control switches together, it greatly reduces the chances that a thief can find them all before the alarm sounds and the pager relays its message.

Installing A Backup Battery - Even if your system does not come with a backup battery it is a good idea to install one. Mount it in the engine compartment or in the trunk making sure the unit or the wires cannot be reached from underneath the car.

Alarm systems consume power, so be careful that your battery, alternator belts and electrical system are in good working order. Find out how much current your alarm system requires so you won't overload the electrical system.

WIRING THE ALARM SYSTEM

It is extremely important to follow the manufacturer's wiring instructions carefully. Make all of your connections *before* connecting the power. Improper wiring connections can damage the control unit and sensors while invalidating the warranty.

Most kits come with wire, but sometimes you will need more than the manufacturer supplies. Always follow the wire gauge recommendations if you have to buy additional wire.

Make sure to connect the power lead to the control unit on the protected side of the fuse box using the proper size fuse. Also make sure this lead is not turned off by the ignition key. Finally, check for proper grounding of the control unit.

Remember to follow the drilling precautions mentioned above. Then drill the holes and run the wires for the

TYPICAL WIRING DIAGRAM

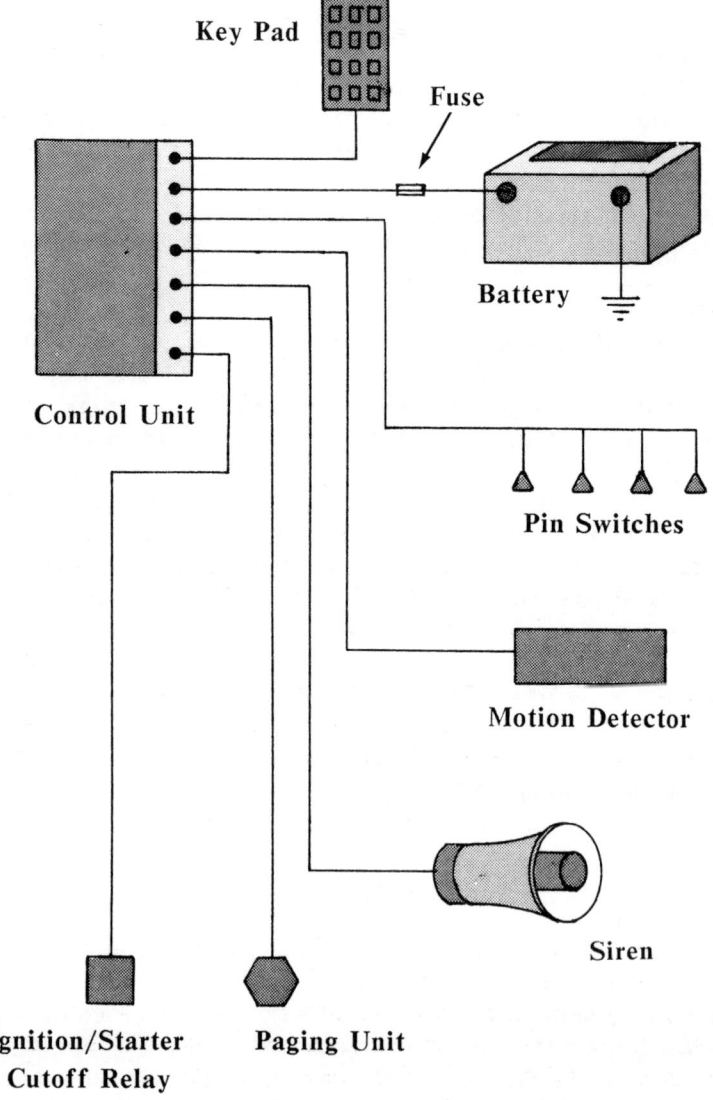

How to Install Your Own Alarm System

sensors, detectors, alarms, transmitters and receivers. When connecting switches and sensors, avoid running any of the wiring where it can be reached by a thief. But if you have no other alternative, protect the wires from tampering by placing them in a metal cable. In addition, make sure that any wiring won't be frayed by running it next to an abrasive surface. Try running the wires next to the vehicle's wiring harness and avoid running them under the carpets if possible.

Also try running wires between the sensors, detectors, switches, alarms and the control unit with as few splices as possible. If you do have to splice wires together, solder the connection and double wrap it in electrical tape. The idea is to keep potential problems in the connections to a minimum.

The final thing is to connect the power and test the system.

TESTING THE SYSTEM

After you have finished installing your alarm system, test it. Check all the switches, sensors and detectors. Set off the system by opening the doors, trunk and hood. Jack up and bump your vehicle to test the motion and shock sensors. Make sure the paging system works and that the siren, bell or horn can be heard at least a block away.

At first, it is best to set all of the sensors on the least sensitive setting and then increase the sensitivity while testing.

Also test your system periodically to insure that it will work if needed. Check all the switches, sensors and alarms as well as the backup battery.

FALSE ALARMS

Hopefully, you have installed your system properly so that false alarms will be minimized. False alarms can be caused by water getting into switches and by transient voltage spikes caused by antenna motors, ignition coils, door locks and from fan motors. They are also caused by detector sensitivity adjustments set too high, faulty wiring, equipment or installation.

Vehicle Alarm Systems

False alarms irritate neighbors and give all alarm systems a bad reputation. But some false alarms are inevitable. Make sure you understand your alarm system thoroughly and remember to disarm it when you return to your car.

VEHICLE ALARM ORDINANCES

Local governments are trying to fight back against the problem of false alarms. Many jurisdictions require that a system automatically reset itself after a prescribed time. Some impose fines and will even tow a car away if the alarm sounds indefinitely. It's a good idea to know what the law is in your area before installing your system.

The next chapter shows you other protections available. Many of these devices are installed to supplement your alarm system, while others offer a great deal of protection in their own right.

ALARM CENTER, INC.
7150 CONVOY COURT, SUITE A
SAN DIEGO, CA 92111
(619) 571-3535

SOLD TO:

Cash Customer

ITEM NUMBER	DESCRIPTION
	ALARM INSTAL BOOK
Received $ 14.00 Cash	

SUBTOTAL		SALES TAX	INVOICE
12.95		0.94	

ALARM CENTER, INC.
7750 CONVOY COURT, SUITE A
SAN DIEGO, CA 92111
(619) 571-3535

SOLD TO:

Cash Customer

ITEM NUMBER		DESCRIPTION	
.		ALARM INSTAL BOOK	
Received $	14.00	Cash	

SUBTOTAL		SALES TAX	INVOIC
12.95		0.94	

INVOICE

INVOICE NUMBER   R-100792
INVOICE DATE   09/17/90
SALESPERSON NO. 3
DUE DATE
DISCOUNT DATE
PAGE   1

QUANTITY	PRICE		EXTENSION
1.00	12.95		12.95

TAL	AMOUNT PAID	BALANCE	CHANGE DUE
.89	14.00	0.00	0.11

INVOICE

INVOICE NUMBER	R-100792
INVOICE DATE	09/17/90
SALESPERSON NO.	3
DUE DATE	
DISCOUNT DATE	
PAGE	1

QUANTITY	PRICE	EXTENSION
1.00	12.95	12.95

TOTAL	AMOUNT PAID	BALANCE	CHANGE DUE
13.89	14.00	0.00	0.11

Chapter 4

Other Vehicle Security Devices

In this chapter you will see other protection devices you can install on your vehicle to protect specific items and to supplement your alarm system.

VEHICLE IMMOBILIZERS

If you decide not to install a complete alarm system in your car or truck, you should at least consider installing something to prevent your vehicle from being started and driven.

In Chapter 2, ignition, fuel and starter cutoffs were described as part of an alarm system. However, you can purchase these devices as separate units.

Ignition Cutoff - This involves installing a hidden switch that can interrupt some portion of the ignition circuit. A simple technique used on non-electronic ignition systems is to install a circuit that grounds the ignition coil on the negative side. Once grounded, the coil will not work. This circuit is controlled by a hidden switch located under the dash, under a seat or in the trunk.

This technique is not recommended, especially on electronic ignitions. It may damage the ignition components causing hundreds of dollars in repairs. *And it may void your new car warranty!*

Another problem with interrupting the ignition in today's vehicles is two fold. First, it may be illegal in your

Vehicle Alarm Systems

state to tamper with the ignition system because of emission control regulations. And second, a malfunction while driving is potentially dangerous. It could cause the engine to quit, making the power steering and power brakes very difficult to use.

If you are still interested in an ignition interrupter, it is a good idea to buy a kit that comes with a wiring schematic for your vehicle or obtain a copy from a shop manual. Make sure you know exactly which wires to cut and splice. Don't experiment at the risk of damaging your ignition system.

Electric Fuel Pump Cutoff - Here is another technique to immobilize your car or truck. The idea is simple, but you do have to have an electric fuel pump. Without the fuel pump working, the car uses up the remaining fuel in the carburetor and then quits. The thief is taken by surprise forcing him to abandon the car.

All that is involved is to install a switch in the electric fuel pump circuit. However, most installers don't recommend trying this on a fuel injected vehicle. Fuel starvation may damage some injection systems.

Don't try fuel starvation with a diesel engine for the same reason. Instead, install a starter motor cutoff.

Starter Motor Cutoff - Interrupting the starter seems to be popular with professional alarm installers. You don't have the problems associated with ignition interrupters but it is possible, but not probable, that a thief could push start the car in order to steal it.

If you install a starter motor cutoff, keep the wires from the switch to the starter as short as possible, because amps are lost as electricity travels through the wiring. Most starters need up to 40 amps, so use a heavy gauge wire that is insulated and protected.

Fuel Cutoff - The three methods described above interrupt an electrical circuit. With a fuel cutoff valve you prevent the fuel from getting to the engine.

A valve is installed in the gas line between the fuel tank and the engine. This valve is controlled either electronically or mechanically, with a hidden switch or with a key.

One of the problems with this method is having to splice into a fuel line. Another involves locating a mechanical valve

Other Vehicle Security Devices

in the passenger compartment. It could allow fuel to spill inside the vehicle.

SECURING HOODS & TRUNKS

You want to secure the hood and trunk for two reasons. Sometimes a thief is only interested in what can be stolen out of the trunk or from under the hood. If you have ever had to replace a carburetor or air conditioning compressor you know how expensive and valuable these parts are. Also consider what you keep in the trunk. Many times we use it to lock packages and other valuables.

The second reason to secure the hood and trunk is to protect the alarm control unit, wiring and warning sirens from being disabled.

Unfortunately, standard hood and trunk latching mechanisms are not secure. For a thief with a crow bar, it is not difficult to pry open either the trunk or hood. And if the thief gets inside the car, he can easily release the hood and sometimes the trunk. I hope you see why additional locks are needed.

One of the simplest methods of securing a hood or trunk is with a case hardened chain and pad lock. But chaining these openings is inconvenient and doesn't stop a determined thief, it only slows him down.

Another method is to install locking pins in the hood or trunk. This is not as awkward as a chain, but it does have its drawbacks.

A more practical solution is to install a dead bolt locking mechanism. A dead bolt hood and trunk lock is operated by an electric solenoid or mechanical cable from the passenger compartment. What makes these locks more secure is that a key is needed to release the dead bolt. This prevents unauthorized access to the engine compartment or trunk which is very important when alarm control units or alarm devices are mounted there.

One problem with solenoid operated dead bolts is if the battery goes dead. The car can't be jump started because the hood can't be raised.

HOOD DEAD BOLT

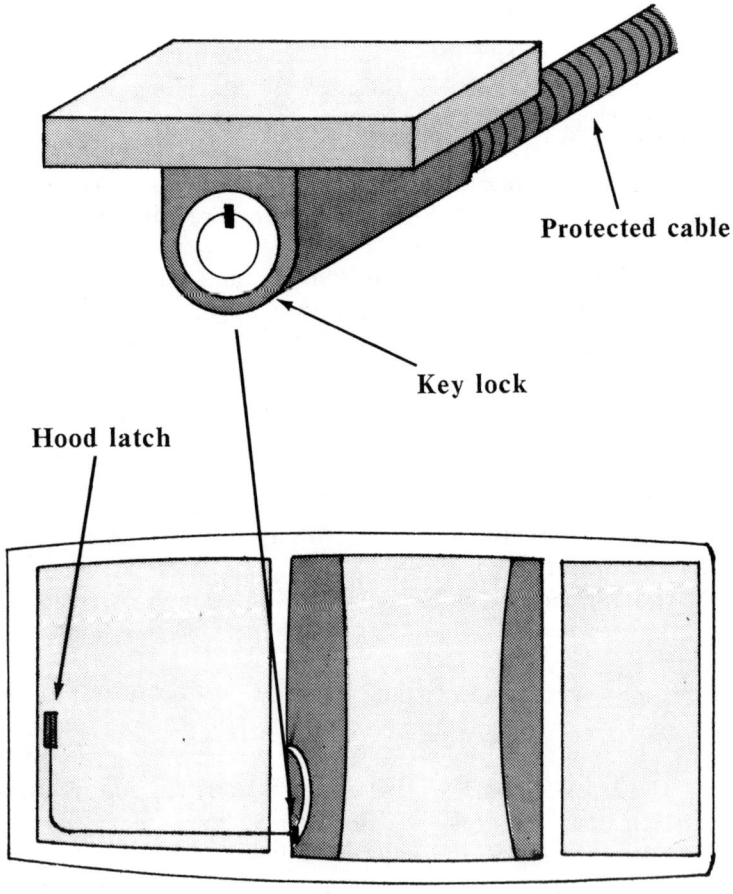

Other Vehicle Security Devices

WINDOW GLASS ETCHING

The technique of etching the vehicle identification number or "VIN", into the window glass has been around for a few years. It has been used successfully in Europe, but really hasn't caught on in the United States.

Security experts say that etching discourages most professional car thieves who steal cars for resale. A pro wants to maximize his profit on each stolen car. When all the windows of a stolen car have to be replaced, it takes additional time and money for the thief to correct.

The idea is to permanently inscribe each window with an identification which can be used to trace the vehicle. In the past, etching was done by sandblasting identification numbers into the windows. However, **Security Etch International** of Newport Beach, CA., has developed a faster and better way of etching vehicle windows. Their method uses an acid paste and precision cut stencils.

For the do-it-yourselfer, **Security Etch International** has developed a kit that is quick and easy to use. But if you don't want to try this yourself, Security Etch is also available through professional alarm installers. If you are serious about protecting your vehicle, window etching is a very cost effective way of deterring professional thieves. It is an economic deterrent to auto theft which compliments the physical deterrent of an alarm system.

PROTECTING RADIOS AND TAPE DECKS

Car radios and tape decks are high theft items and need protecting. You should also keep a record of the model and serial numbers of your unit. Mark them with your own identifying number such as your drivers license or vehicle identification number. Without identifying numbers that can be traced back to you, the police have no way of returning your property if it is stolen and then later recovered.

There are two ways to give your sound system additional protection. One manufacturer has come up with a key locked panel which fits over your radio or tape deck. This prevents a thief from unfastening the two nuts which hold the

PROTECTING CAR RADIOS AND TAPE DECKS

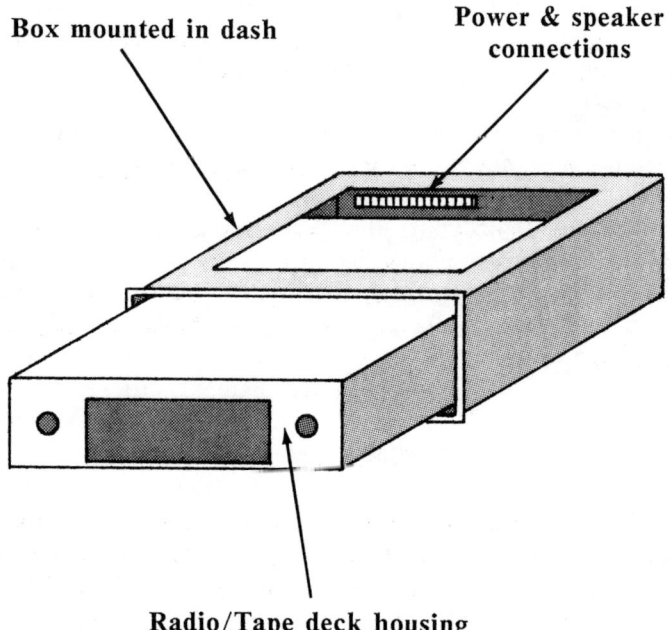

Box mounted in dash

Power & speaker connections

Radio/Tape deck housing

Other Vehicle Security Devices

unit into the dash. But if a thief wants your radio or tape deck badly enough he may try prying this cover off. And if your radio is mounted in a dash or console made of plastic, it is not difficult to completely rip out the radio without disturbing the locking panel. This will be even more expensive to fix than just replacing a stolen tape deck.

Another manufacturer has devised a housing to fit over your radio/tape deck. A matching box is then installed in the dashboard where your radio once was. The housing has a female connector plug which lines up with a male connector in the box. This arrangement allows you to remove your radio or tape deck and take it with you.

If you install one of these security housings, a thief might think you have simply put the radio under the seat or in the trunk and may try breaking in anyway.

PROTECTING WHEELS AND TIRES

Remember the last time you had to replace your tires and how much you had to spend? You might give serious thought about securing tires and wheels with locks. This is especially important if you have high performance radial tires mounted on custom wheels.

LOCKING GAS CAPS

Locking gas caps prevent someone from stealing your gasoline, but more important, they prevent a vandal from putting a foreign substance into your fuel system. Losing a tank of gas is bad enough, but having to flush out the fuel system and rebuild the engine is worse. Installing a locking gas cap is inexpensive when you consider the alternatives.

ALARM WARNING WINDOW STICKERS

One way to discourage a car thief is to place warning stickers on your windows. Some car owners use the stickers even if they do not have an alarm system. Many times the sticker alone are enough to discourage a thief.

Vehicle Alarm Systems

IMITATION KEY PADS

Imitation key pads installed on the dash board or console will also discourage a car thief. The idea is to make him think your car is protected by an alarm system. Key pads are available that look authentic.

STEERING WHEEL LOCKS

Another method to discourage a car thief is to use a bar which locks the steering wheel to the brake pedal. One end of the bar goes around the pedal while the other locks around the wheel. This prevents the car from being driven because the steering wheel cannot be turned and the brakes cannot be applied. This may deter an amateur from taking your car, but it will not stop a pro.

ANTI-THEFT IGNITION LOCKS

Chapter 1 describes how easy it is for a pro to extract an ignition lock and start a car. One way to prevent this is to install a locking device which encircles the steering column and ignition switch.

This prevents a thief from extracting the lock and driving off with your car.

OTHER PRECAUTIONS AND CONSIDERATIONS

Vehicle Registration - Don't leave your registration card in your vehicle. It is better to carry a copy in your wallet or purse. If a thief were to steal your car he would know who you are and where you live. This is especially important when you park in a long-term lot at an airport.

Parking Your Car - How and where you park your car is also important in deterring a thief. You want to prevent a

Other Vehicle Security Devices

thief from having access to your vehicle where he can break into it without being seen.

If you have to park on the street, don't park on the end of a block. It is better to park between other cars in the middle of the block to prevent a thief from simply backing up a tow truck and hauling your car away.

Park in different places from day to day. Don't make a pattern of your parking habits. A thief will spot your routine and know the best time to strike.

Park where your car is visible. Put it on a busy street so a thief won't have privacy while trying to break into your car. At night, park your vehicle under a street light for the same reason.

Keys and Credit Cards - Always take your keys out of the ignition. Make it a habit every time you get out of your car and never get out and leave the engine running. Not only is this dangerous if the car slips into gear, but an unattended car with the motor running is an irresistible opportunity for any would-be car thief.

Don't hide a key under the hood unless it is very well hidden. Car thieves usually know where to look. It is a better idea to carry a spare in your wallet or purse.

When you have to leave your car at a parking lot or for service, leave only the ignition key and never your house or office keys. Who knows how many people will have access to your keys. It would be very easy for a dishonest employee to make duplicates giving him access to your home, office and possessions.

Finally, never leave credit cards or identification papers in your car. There is no reason to make a thief's job any easier.

Chapter 5

Professionally Installed Alarms

Hopefully, you have purchased and are reading this book because you are interested in finding out how to prevent a thief from stealing your car and its accessories.

The worst time to buy a security system is right after you have been ripped off (that is if you are lucky enough to get your vehicle back!) You're upset and feel vulnerable. Some uncaring criminal has disrupted your life while costing you time and money. The only thing you know is that you don't want it to happen again - <u>ever</u>!

It is unfortunate that many alarm systems are installed *after* a thief has stolen the tape deck or the car itself. There is a big temptation to fix the problem as soon as possible by installing something.

So what should you do? Do you run right out and buy the first security system you see? Of course not. This chapter gives you a more rational approach to this problem.

EVALUATING YOUR NEEDS

The first thing to do is to determine just what type of a system you should install. For example, do you need a paging system or will a siren or horn be adequate to protect your car? Do you want a transmitter to arm and disarm your system or will a hidden toggle switch suffice?

The key to determining what system to install and how much to spend depends on where you drive and where you park your vehicle. If you lock it in an enclosed garage every night and you are the only one to have access to it, maybe you could get by with a simple alarm system. This system would include a warning siren and ignition cutoff for those times you have to park it in an unprotected area. Obviously you will need more protection if you drive and park your car in high crime areas and leave it unattended for long periods. Here, the most sophisticated system with multiple sirens, a paging system and ignition or starter motor cutoff would probably be best.

Take a few moments to think about how you park your car. At home, do you have a private garage, or do you have to park on the street, in an unattended lot or in an open carport? At work, do you park in a guarded parking structure or could a thief break into your car without being noticed? The time you take now to think about your own situation can prevent you from buying a system that won't protect your car properly while giving you a false sense of security.

If you over estimate this threat, you can easily spend far too much money for your alarm system. You have to be especially careful of this when talking to an unscrupulous alarm dealer, especially if you have just had your car stolen or broken into. At this time you might be easily talked into buying something you don't need.

An alarm system minimizes your chances of having your car burglarized or stolen. However, remember that a skilled and determined thief can probably steal your vehicle if he really wants it. Your alarm system should be efficient enough to not only slow the thief down, but deter him completely.

Ok, now that you have thought about where and how you park your car, here is a summary of alarm options:

    The Control Panel
        Lock Up or Latching Relays
        Instant Circuits
        Delay Circuits
        Additional Circuits
        Override or Valet Switch
        Automatic Reset
        Ignition System Cutoff
        Starter Motor Cutoff
        Electric Fuel Pump Cutoff
        Backup Battery

Professionally Installed Alarms

Switches, Sensors and Detectors
- Pin Switches
- Motion or Shock Detectors
- Voltage Drop Sensors
- Ultrasonic Sensors
- Sound Discriminators
- Panic Switches

Alarms
- Sirens
- Bells
- Car Horns
- Air Horns
- Voice Synthesizers
- Flashing Headlights

Paging Systems
- High Gain Antennas

Arming and Disarming Devices
- Passive Arming
- Hidden Toggle Switches
- Key Operated Switches
- Digital Key Pads
- Radio Transmitters

When buying a protection system for your car, you have many options from which to choose. The best system for you is one that is reliable, simple to use and protects your vehicle based upon your individual needs. Be sure to consider how you want to turn the system on and off (passive or active arming); what will trigger an alarm (switches, sensors and detectors); and what alarm devices will be activated (sirens, horns, bells, headlights and paging systems).

Another important thing to do is to call your insurance agent or broker. Check to see if your insurance carrier offers discounts for installing an alarm system. (Many auto insurance companies offer up to 15% off of your premium.) Find out what type of system they recommend, if any, and ask what other requirements the system must meet to receive the discount.

Vehicle Alarm Systems

HOW TO FIND A PROFESSIONAL INSTALLER

Most vehicle alarm systems are sold and installed through new car dealers, car radio dealers and through shops specializing in vehicle security systems. There are franchised dealers who only install specific makes, while some installers put together custom designed alarm systems from various alarm component manufacturers.

The trick is to find an installer who knows what he is doing. (This is usually easier said than done!) You want someone who knows and understands alarm system design and installation, preferably one who specializes in vehicle alarm systems.

The best place to start is in the yellow pages of the telephone book. Look under the headings of *"Burglar Alarm Systems - Automobiles"* or under *"Automotive Security Systems"*.

If you are in the process of buying a car, most new car dealers can either install a system, subcontract the job or refer you to a qualified alarm installer.

One of the best sources is a referral from a satisfied customer. Ask your friends, neighbors and co-workers who have had alarm systems installed on their vehicles. Find out what their experience has been with the dealer, the installer and the equipment. Ask if they have had problems with false alarms or getting the system serviced.

HOW TO CHOOSE AN INSTALLER

After you have assembled your list of auto alarm installers, you should find out if the company is reputable. Do this by contacting the Better Business Bureau and check to see if the company you are considering has had any complaints filed against it. Also check with your local consumer protection agencies for similar complaints.

Always be cautious when dealing with any new installation company. A new company has not had time to develop a track record, whether it is good or bad. Dealing only with established alarm companies with numerous satisfied customers makes good sense. A new company may offer the lowest price or best guaranty, but what good is a guaranty if they are out

Professionally Installed Alarms

of business in six months and you have a problem. I think you get the point.

Here is a list of questions to ask the alarm company or installer you are considering:

1) How long has the company been in business?

2) How many systems have they installed?

3) Have their installers been screened for criminal records?

4) Are their installers bonded?

5) Do they have the proper city, county and state licenses and permits?

6) Can they give you names of satisfied customers?

7) How long will you have to wait before they can install your alarm system?

8) How long with the installation take?

9) How long will the alarm company guaranty the entire alarm system after the installation is complete?

10) Will the company finance the system or must you obtain your own financing?

11) What liability does the alarm company assume if the alarm system malfunctions and the vehicle owner suffers a loss?

Make sure you have a clear understanding with the alarm company and the installers before signing a contract. Review their liability limitations and liquidated damages clause. (Alarm companies try to limit their liability by using such a clause.) Also, liability limitations differ from state to

Vehicle Alarm Systems

state. Try to remember that you are buying an alarm system and *not* insurance against loss.

In addition to checking out the installers, you also want to find out about the specific alarm systems they install. Ask these questions:

1) What brands of alarm equipment do they install?

2) How long has the manufacturer(s) been in business?

3) What has been the alarm company's experience with the manufacturer and the reliability of the alarm equipment?

4) How long has the alarm company installed this brand of equipment?

5) How long is the warranty and what is covered: parts *and* labor?

6) Who services the components and how long will it take if repairs have to be made?

7) Can they guaranty that the installation will not void your new car warranty?

Finally, when your new alarm system is finished, have the installer show you how the various switches and settings work. Even though they have probably tested the system before you pick up your car, have them test it again while you watch. The point is to make sure everything is working *before* you drive away.

It is also a good idea to test all of the electrical equipment that is not related to the alarm system, such as power windows, door locks and sun roof. If anything should malfunction after the alarm is installed, you want to find out now if the installers had anything to do with it. You might have trouble later getting the installers to accept responsibility after your vehicle leaves their shop.

PART II

Home Security Systems

Dedication – Part II

To the burglars who violated the privacy of my home and showed me how vulnerable I was.

Chapter 6

Protecting Your Home & Family

Auto security systems share many of the same principles of detecting a break-in and relaying an alarm message used in home alarm systems. The sensors, panels and alarms are very similar in design.

Vehicle alarms are designed primarily to protect personal property such as a stereo, cellular car phone or the car itself. But home security systems have a much bigger job to do. By contrast, the primary purpose of a home alarm system is to protect the life of your family. Protecting possessions is secondary.

BASIC PROTECTION SYSTEMS

Home security systems use two basic types of protection to detect burglars and intruders: perimeter protection and area protection. A perimeter protection system surrounds your home with security by protecting every point where an intruder could gain entry. In contrast to perimeter protection, an area protection system guards the interior of your home by detecting the intruder's presence in a room or space.

Your home security system can do much more than protect against burglars and intruders. Any security system worth installing should protect your family with special protection systems as well. These special protection systems will sound a warning and summon help in the event of a fire, medical emergency or other life threatening situation.

Home Security Systems

To understand how a home security system works, let's take a closer look at these protections.

Perimeter Protection

Perimeter protection is the first line of defense in a home security system. It detects burglars and intruders as soon as they try to break into a house. Sometimes this is called *point of entry* protection.

When activated, a perimeter system senses when a door or window is opened, broken or rattled. Sensors and switches are installed at every point a burglar could get into your home. For example, doors and windows are usually protected with a magnetic or plunger switch. In addition to these switches, many homeowners also install specially wired window screens as backup perimeter protection.

Switches, sensors and screens on doors and windows provide only partial protection against an intruder. What happens if a burglar or intruder simply smashes the glass in a door or window? He could then climb through the opening without activating one of the switches. For perimeter protection to be effective, you should also consider installing glass breakage sensors to initiate an alarm if a window or door glass is cracked or broken.

Some brands of alarm systems protect both doors and windows with a sensor called **a sound discriminator. A sound discriminator** will "listen" for glass breaking or wood splintering - the sounds an intruder makes as he breaks into a home.

For perimeter protection to be effective, every potential entry point should be protected with a sensor or switch. This would include all doors, windows, basement doors, basement windows, sliding glass doors and skylights.

Area Protection

Area protection is the second line of defense against burglars and intruders. Instead of detecting a door or window opening, or glass breaking, area protection systems detect the presence of an intruder *after* he is inside the house. This protection is sometimes referred to as space or interior protection.

Area protection sensors and detectors are more sophisticated and usually more expensive than their perimeter protection counter parts. They are usually placed where a burglar is most likely to pass through as he searches for valuables. Hallways, stairways, and the master bedroom are potential installation locations. Microwave, ultrasonic, passive infrared, photoelectric beam and pressure sensitive mats are the most popular of these area sensors.

Combining Area & Perimeter Protection

Some homeowners will install only one type of protection system, but you will be much safer by installing both perimeter and area protection systems. The idea is that one system backs up the other if the equipment fails or if an intruder defeats part of the alarm system. The best home security systems use this combination of protection.

Obviously this dual system will cost more, so when deciding upon a protection system or systems for your home, take into consideration what possessions you have to protect and how much you can afford to spend. (Determining your security needs is explained in Chapter 8.)

SPECIAL PROTECTION SYSTEMS

Alarm systems can be used for more than detecting a burglar or intruder breaking into your home. Special protection switches and sensors can be added, enabling you to tailor your system to your specific security and safety needs. The most popular special protection systems include smoke and heat detectors as well as emergency panic switches.

Smoke and Heat Detectors add another dimension of safety to your alarm system. These sensors and detectors warn you to get out of a burning house before it is too late. Many alarm systems on the market today accept these fire protection sensors giving you and your family an extra margin of protection.

Home Security Systems

Emergency Panic Switches are popular with homeowners because they initiate an alarm instantly when the button is pressed. Their primary purpose is to provide protection if an intruder tries to break-in while you are inside your home. It is a good idea to install one within reach of entry doors and in the master bedroom. Hand-held transmitters are also available which can trigger the emergency panic circuit and sound the alarm from anywhere in your home.

In addition to sounding the alarm, emergency panic switches can be installed to turn on inside and outside lights. Many times an alarm siren combined with lights turning on is enough to frighten away most burglars.

Medical Switches are essential to some homeowners with special health problems. If you or some member of your family becomes unable to dial the telephone but is still capable of pressing a button, medical help can be summoned quickly. Your alarm system's telephone dialer can call your doctor directly or a central station can dispatch paramedics.

Other Protection Systems are available for your security system which detect dangerous levels of propane, natural gas or carbon monoxide in your home. Some will even detect water flooding your basement. What protection system you decide to install depends on your own special requirements.

HOW AN ALARM SYSTEM WORKS
BASIC COMPONENTS

First an alarm system must detect the emergency whether it is intrusion, fire or medical. Once a sensor or switch is activated, it sends an alarm message to the master control panel. This control panel interprets the message or signal and initiates an alarm to summon help. As you have already seen, it does this with perimeter, area and special protection systems.

Switches, Sensors & Detectors serve different purposes: they detect burglars and intruders, and they sense smoke and

HOW AN ALARM SYSTEM WORKS

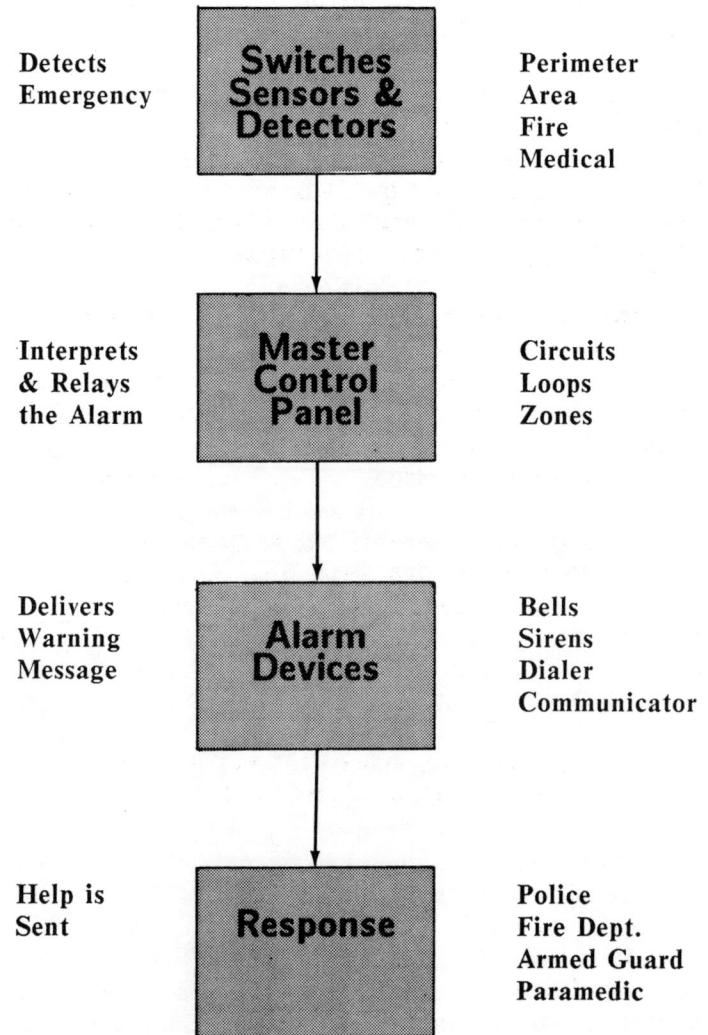

Home Security Systems

heat. Their job is to set in motion the process by which help is sent.

The Master Control Panel is the nerve center of the alarm system. It receives alarm messages from the switches, sensors and detectors, interprets them and sends for help via the local alarms, the automatic telephone dialer and the digital communicator.

Alarms are used to warn others of the emergency and request help. Their purpose it to attract attention, send taped voice messages and relay electronically encoded emergency information. Most home alarm systems do this with sirens, bells, automatic telephone dialers and digital communicators.

The alarm sirens and bells are usually installed both inside and outside the house. They scare an intruder and warn the neighbors to call police.

Automatic telephone dialers dial preprogrammed telephone numbers and then deliver a prerecorded tape message when a connection is made.

Digital communicators send alarm messages to a central monitoring station at an off site location. When a message is received, the monitor at this location is the person responsible for requesting the appropriate help.

HOW ALARM SIGNALS ARE TRANSMITTED

Alarm signals and messages are relayed from the sensors and switches to the master control panel by three methods. An alarm system will use hard-wiring, wireless transmitters and line carriers. Some systems will use only one of these methods while others will use a combination of two or all three.

Hard-wiring connects switches and sensors to the master control panel by running wires through walls, under the house, in the attic and behind trim molding. Hard-wiring an alarm system can take a lot of time and be difficult for both the professional installer and do-it-yourselfer.

Wireless Transmitters send these same alarm messages to the master control panel by radio waves. First, switches and sensors are mounted and then connected to a transmitter. When the control panel is equipped with a compatible receiver, alarm messages can be relayed.

Alarm system installation is faster but sometimes more expensive with wireless transmitters. However, for the do-it-yourselfer the extra cost might be money well spent. It is sometimes difficult and time consuming to run wires between switches, sensors and the master control panel because of restricted access in the basement, crawl space or in the attic. Drilling holes and pulling the connecting wires through walls is also tedious work.

Professional installers will sometimes use wireless transmitters to save on labor costs. The extra expense for the transmitters and receiver can sometimes save you money if you have your system installed professionally.

On the negative side, these transmitters are usually battery powered and if the batteries fail, the alarm system is vulnerable to an intruder. If you install a wireless system it is important to check the transmitter batteries at least once a month and replace them at least once a year.

Line Carriers, sometimes called carrier currents, use your existing house wiring to send and receive alarm messages. They do this by first receiving an alarm message from a switch or sensor. After encoding the message, it is sent to the master control panel via the house's electrical system.

Alarm messages can be sent and received when the master control panel is plugged into another outlet on the same wiring system. These systems are often called "plug in" systems because you just plug the sensors into electrical outlets. Because of the easy installation, line carriers are popular with do-it-yourselfers.

HOW EMERGENCY HELP IS SUMMONED

After the master control panel has received an alarm message from a switch or sensor, the control panel must relay this alarm message notifying someone to call for the police, fire or medical help.

Home Security Systems

Local alarm sirens and bells warn your neighbors to call for help, but sometimes the siren or bell might not be heard and no help will be summoned.

Automatic telephone dialers and digital communicators send their alarm messages to remote locations over regular telephone lines. The person receiving the taped message from the automatic telephone dialer can request the police or fire department response. But with a digital communicator, a central station monitor can receive more detailed information about the emergency. He can determine where in the house the intruder is or tell exactly where a fire is burning before requesting help.

Radio transmitters are also used to send alarm messages from the master control panel to the central station. They work on the same principles used by the wireless transmitters described above but are much more expensive and are rarely used in most home installations. On the other hand, these transmitters offer high levels of security because they are not subject to service interruptions that can affect automatic telephone dialers and digital communicators. In addition, they can be used to communicate to a central station from a location where no telephone lines exist.

One of the most recent developments in home security monitoring is two-way cable television. Most homeowners are familiar with cable television, but in some communities this same cable can also be used to send alarm information to a central station. Hopefully, alarm systems monitored by cable operators will expand to more communities while making a monitored system more affordable.

Chapter 7

Alarm Components & How They Work

In the previous chapter you saw the basics of a home alarm system: the types of protection, the basic components, and the methods used to send and receive alarm messages. If you have stayed with me thus far, you should have some understanding of home alarm systems. Before we get to the planning and installation, we must take a closer look at the components that make up an alarm system. This chapter describes and explains the various components you may encounter.

THE MASTER CONTROL PANEL

The master control panel is the nerve center of an alarm system. Its primary functions are to receive messages from switches and sensors, interpret what type of alarm condition exists, and then relay the appropriate alarm message to summon help.

Master control panels can be very basic or quite complex. The simplest control panels provide only one or two protective circuits, a connection for a local alarm, and a few other features. The number of switches and sensors the panel will accept is limited. These control panels are used for minimum security systems in small houses or apartments.

By contrast, the more sophisticated control panel provides four or more protective circuits or loops. A separate fire protection circuit is usually included for smoke and heat

THE MASTER CONTROL PANEL

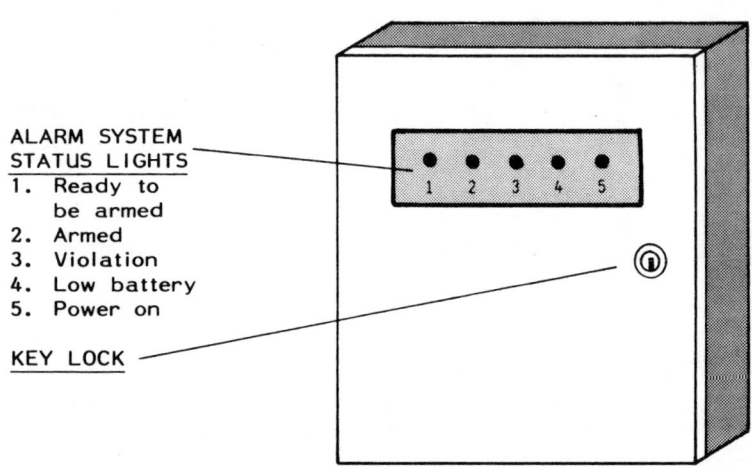

ALARM SYSTEM
STATUS LIGHTS
1. Ready to be armed
2. Armed
3. Violation
4. Low battery
5. Power on

KEY LOCK

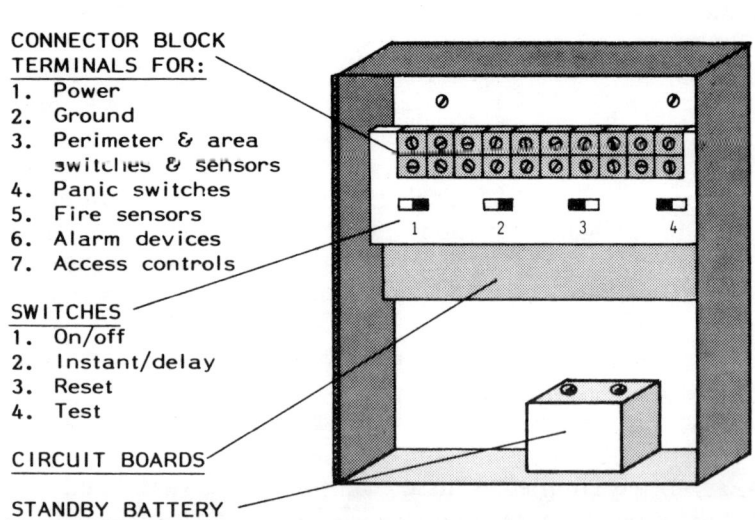

CONNECTOR BLOCK
TERMINALS FOR:
1. Power
2. Ground
3. Perimeter & area switches & sensors
4. Panic switches
5. Fire sensors
6. Alarm devices
7. Access controls

SWITCHES
1. On/off
2. Instant/delay
3. Reset
4. Test

CIRCUIT BOARDS

STANDBY BATTERY

Alarm Components & How They Work

detectors. These sophisticated panels have connections for automatic telephone dialers and digital communicators as well as connections for local alarm sirens and bells.

Some panels have light emitting diodes (LED's) which tell you if the system is armed or disarmed. In addition, the LED's indicate if the zones, circuits or loops are functioning normally. Status lights can also show if the standby battery is properly charged.

These sophisticated panels are used in larger homes because more protective circuits are needed as the total number of switches and sensors increases. Also, they provide more security features for higher levels of protection.

Combination Panels - Between the basic and sophisticated panels are master control panels which combine an area sensor such as a microwave, ultrasonic or sound discriminator with the panel. These combination units are popular with homeowners who install their own alarm system. But there is a compromise. You save money on equipment while giving up system versatility. It may be difficult to find a combination unit that has the features you need in a control panel as well as an area sensor which fits your security requirements. Also, it is difficult to upgrade an alarm system with a combination unit without replacing the control panel.

Chapter 8 will show you how to choose a master control panel to fit your needs. Before we get there, let's look more closely at how the master control panel works and what it can do.

HOW THE CONTROL PANEL RECEIVES MESSAGES

Different circuits or loops are used in the master control panel for intrusion, fire, panic and medical emergencies. Circuits used for perimeter switches and sensors are usually different from those used for area protection sensors.

By using multiple circuits, the master control panel can tell what type of emergency is taking place and where it is located. Multiple alarm circuits are used to "zone" an alarm system. Zoning divides the house into areas of protection or separate reporting zones. Each area or zone will have its own protection circuit or loop. By connecting switches and sensors by function and by zone, the master control panel can locate and identify the emergency for a central station monitor. Police, fire or medical help can then be sent.

Separate circuits and zones are more expensive to install but they do provide more security. They are also important if your alarm system malfunctions and needs service. Being able to locate the trouble with a zoned system, repairs can be made faster saving you money and getting your system back into operation promptly.

CIRCUIT CHARACTERISTICS OF CONTROL PANELS

A circuit or loop is referred to as being either normally open (N.O.) or normally closed (N.C.). Sometimes you will see these circuits referred to as O.C. for an open circuit and C.C. for a closed circuit.

Normally Open - No electrical current is present in this circuit under non-alarm conditions. Sensors and switches are connected to it in parallel. When a sensor or switch connected to a normally open circuit is tripped by a burglar, contacts in the switch or sensor are closed completing the circuit. This circuit is now able to signal the master control panel to relay the alarm message.

Normally open circuits have the advantage of using no electrical current until a contact is closed. But this can be a disadvantage if a burglar cuts the wires of a N.O. circuit. The intruder has compromised the alarm system and the alarm message will not be sent. If a wire or connection breaks or malfunctions, a N.O. circuit cannot send an alarm message to the control panel. Normally open circuits are used where wires are protected and where a burglar or intruder cannot get to them.

Some N.O. circuits can be monitored by the master control panel to determine if the circuit is operating properly. These are known as supervised circuits and are discussed below.

CIRCUIT CHARACTERISTICS

Normally Open Circuit (N.O.)
(in a non-alarm condition)

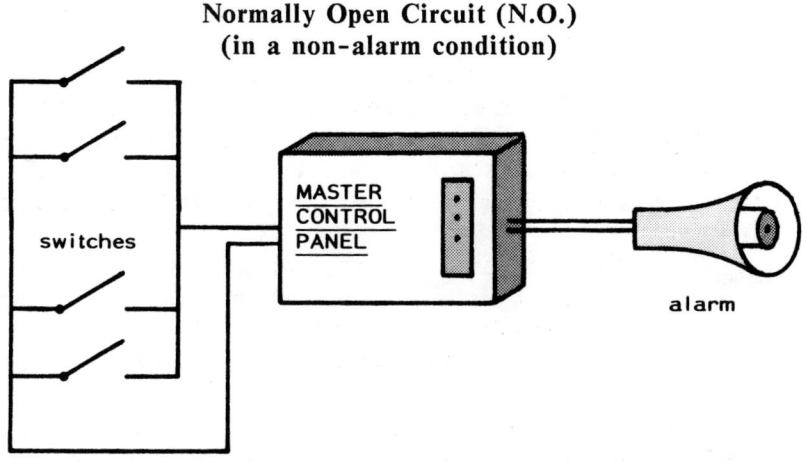

Normally Closed Circuit (N.C.)
(in a non-alarm condition)

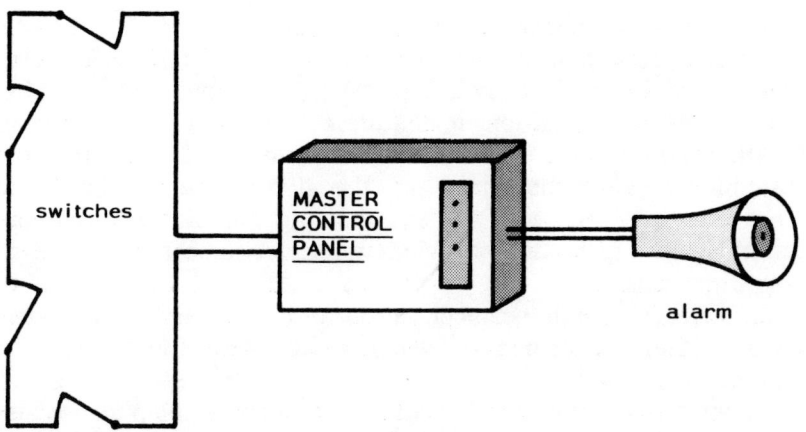

Home Security Systems

Normally Closed - In this circuit, switches and sensors are connected to it in series so that the same current exists in each switch or sensor. Any break or interruption in the circuit signals the master control panel.

Normally closed circuits use small amounts of electrical current in non-alarm conditions. A switch or sensor detecting an open door or window, or broken glass, will break the N.C. circuit. When the master control panel senses this drop in line voltage, alarm messages are relayed. The advantage of a N.C. circuit is that if a burglar tries to cut the wires, an alarm is initiated.

Remember that different types of sensors and switches use different types of circuits. In most alarm systems perimeter switches use N.C. circuits while area sensors typically require N.O. circuits. Fire protection sensors such as smoke and heat detectors use a supervised 24 hour N.O. circuit.

Some switches and sensors can be modified to work with either a N.O. or N.C. circuit. However, you will probably want to purchase a master control panel with both types to handle a variety of switches, sensors and detectors.

OTHER FEATURES FOUND IN CONTROL PANELS

Lockup Relays maintain the alarm condition in the master control panel after the signal is received from a switch or sensor. Even if the switch or sensor is immediately restored to a non-alarm position or condition, the master control panel will continue to sound the local alarms as it activates the automatic telephone dialer or the digital communicator. It prevents an intruder from shutting off the alarm system by simply closing the door or window he just came through.

Reset Switches cancel an alarm and reactivate the alarm system into an armed or ready mode. With this switch the homeowner can abort the alarm if he accidentally trips the system. Some systems will abort the automatic telephone dialer and the digital communicator as well. If the system does not abort, you will have to call the central station and give an identifying code indicating that the alarm was false and it was you, the homeowner, who shut off the alarm.

Alarm Components & How They Work

Automatic Shutoff is required by law in many cities. The alarm system must shut down after a specific period to prevent the local alarm bells or sirens from sounding indefinitely.

Automatic Reset reactivates the alarm system after the local alarms have sounded for a given time period and the telephone dialer or digital communicator have sent their messages. The lockup relay maintains the alarm condition, the automatic shutoff turns off the alarm and the automatic reset prepares the alarm system to detect another intrusion.

Alarm Status Lights are used on some of the more sophisticated master control panels. These light emitting diodes (LED's) tell you if the alarm system is armed, ready to be armed or in a test mode. They can also indicate if a door or window needs to be closed before the system can be armed.

Annunciators are alarm system monitoring devices which also use LED's to indicate the status of a circuit, zone or detector. Normally, a green light is used to show that the system is ready to be armed (fail-safe arming indicator) and a red light usually means the system is armed and waiting to detect an intrusion or fire. When an alarm condition is reported, the LED for that zone is lit indicating where the alarm condition occurred.

Delay Entry/Exit Circuits give you time to activate your alarm system when leaving and time to deactivate it upon returning. You first arm the system at the master control panel or from an inside access panel and then exit through the protected door. The protective circuit used for the switch on the door will not initiate an alarm during the delay period. When you return, the entry delay gives you time to enter and shut off the alarm system before an alarm is sent or sounded. Some delay circuits are adjustable giving you from five seconds to two minutes to arm or disarm the system before an alarm message is sent.

Instant/Delay Switches let you select a delayed or instant alarm for the same protective circuit. Use the delay during the day to enter and exit. At night, use the instant

Home Security Systems

mode when you are in the house. If an intruder tries to break-in while you are in the house, the alarm will sound immediately.

Day/Night Circuits are found in some master control panels. These circuits allow you to disarm certain interior sensors and zones. By flipping a switch on the control panel you can move throughout the house and still maintain the protection of the perimeter protection system. This feature is especially useful at night when you are in your home.

24 Hour Circuits function as long as the master control panel has power, either from the transformed 120 volt household current or from standby batteries. Other protection circuits can be disarmed but the 24 hour circuits are always active.

These circuits are used for smoke and heat detectors as well as for emergency panic switches and medical buttons. Being on continuously, 24 hour circuits prevent you from accidentally shutting off a protection circuit that might save your life.

Fire Protection Circuits - Some master control panels have these specialized protection circuits. Only smoke and heat detectors should be connected to them. Fire protection circuits are 24 hour circuits and should be supervised by the control panel. The output from a fire circuit initiates a different alarm device or message which indicates to neighbors or a central station monitor that the emergency is a fire rather than an intrusion or burglary.

Supervised Circuits - These circuits are monitored by an electrical circuit or radio path which sends information periodically on the status of a sensor, zone, loop or power supply. If the sensor, switch or circuit will not operate because of a malfunction in the system, the supervisory circuit warns the homeowner or central station of the failure.

Supervised circuits are use primarily for fire protection circuits but can also used for area, perimeter or other protective circuits.

Being able to check on the condition of your alarm system adds to your safety and security.

Alarm Components & How They Work

Fail Safe Arming - Master control panels will only arm properly if all circuits and sensors are in a non-alarm condition when you try to arm your alarm system. This feature prevents false alarms. Without it some panels might not arm at all if a switch is left open.

Power For Sensors - Most remote sensors require an outside power source. Some master control panels can provide for these sensors as well.

BATTERY BACKUP - STANDBY POWER SOURCES

For maximum protection, your alarm system should have a battery backup system. This extra power source keeps the alarm system operating in the event of a power failure. But more important, it prevents a burglar from rendering your alarm system inoperative by simply turning off the electrical power to your home.

Rechargeable and non-rechargeable batteries are available to power the master control panel, the sensors and the alarm devices if necessary. Standby batteries such as rechargeable nickel cadmium and gel-types will require a battery charger to maintain them at full power.

Nickel cadmium batteries, called "nicads", will deteriorate over time. If your master control panel is equipped with a nicad backup system, it is important to check on the battery's standing time. (Standing time is the time period the battery can power the alarm system by itself.) However, both rechargeable and non-rechargeable batteries must be tested and replaced periodically.

ACCESS PANELS

Access panels give you control of your alarm system at the doors you use to enter and exit your home. They make using your system easier because you can control the system from a convenient location such as from a frequently used door. Without a remote panel, all arming and disarming would have to be done at the master control panel which is usually hidden in an out of the way location.

ACCESS PANELS
for arming/disarming and shunting

DIGITAL KEY PAD

Alarm system status lights for indicating:
1. Armed/disarmed
2. Loop status
3. Shunt status

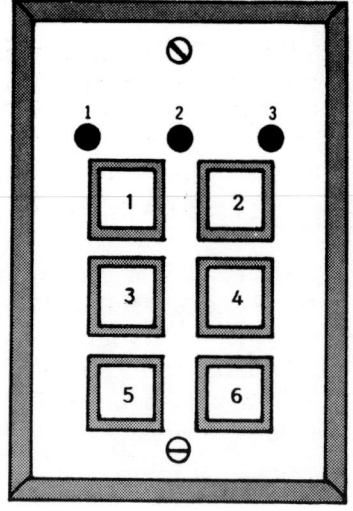

KEY OPERATED SWITCH

Alarm system status lights indicating:
1. Armed/disarmed or shunt
2. Loop status

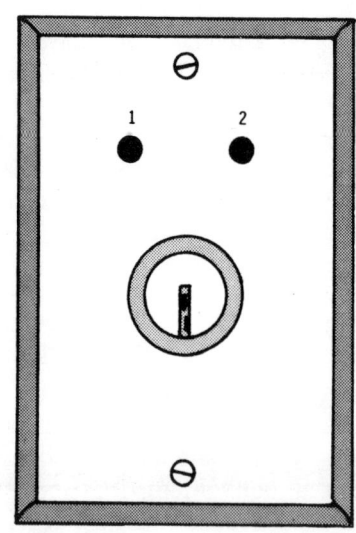

Alarm Components & How They Work

Access panels can be operated with a key or with a digital key pad. The digital pad is popular because it eliminates the need to carry an extra key. In addition, some digital key pads let you select and change your own access codes.

The access panels can be mounted either inside or outside your entry doors. With an inside panel, you will need to use a delay circuit for the door's protective switch. For an outside panel you don't need the delay circuit, but you will need a panel designed for outdoor use. This outside panel is a potential weakness in the security system because a burglar might try to manipulate the access panel to gain entry without setting off the alarm.

To prevent a burglar from compromising the alarm system, outside panels should be equipped with a tamper switch. This switch initiates an alarm if a burglar tries to remove the face plate to the panel and get to the wires running to the master control panel.

Some of the more sophisticated access panels have status lights (LED's) to indicate if the alarm system is armed or disarmed as well as to report on loop and shunt status. These lights help you use your alarm system by reminding you to arm the system when leaving and to disarm it upon returning.

Shunt Switches - These switches are used to deactivate certain sensors, switches or zones. Sometimes you will want to open a particular door or window without having to go to the master control panel or an access panel to disarm the system. You can install shunt switches next to these doors and windows for convenience. As with access panels, shunt switches can be installed either inside or outside your home.

However, outside shunt switches should have a pick resistant lock and a tamper switch.

Shunt switches are sometimes used in an alarm system when cost is a factor or if the homeowner simply does not need the features or controls of the access panels.

SWITCHES & SENSORS FOR PERIMETER PROTECTION

Magnetic Switches These switches are popular because of their reliability and ease of installation. Used primarily for

MAGNETIC SWITCHES

Recessed/Concealed

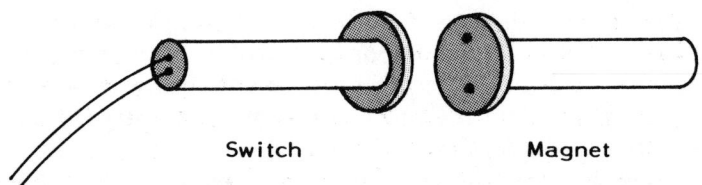

Surface Mounted

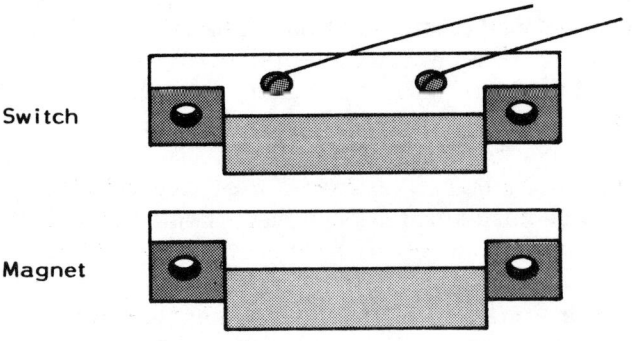

Alarm Components & How They Work

doors and windows, a magnetic switch consists of two separate pieces. One piece is a magnetically operated switch and the other is a magnet. The switch portion is mounted on the stationary door or window frame while the magnet is attached to the door or movable window sash.

When mounted adjacent to each other, the magnet holds the switch in an open position for a N.O. switch and in a closed position for a N.C. switch. If a door or window that is protected with a magnetic switch is opened, the magnet is removed from the switch and the alarm circuit is activated.

Both normally open (N.O.) and normally closed (N.C.) magnetic switches are available, but most perimeter systems use normally closed switches and circuits.

Sometimes you will see the term "reed switch". A reed switch is a magnetic switch that uses two thin metal pieces or "reeds" as contact points. A magnet holds these reeds in an open or closed non-alarm position.

Magnetic switches are of two basic types: either surface mounted or flush mounted. Surface mounted magnetic switches have been installed for years by professional alarm companies because they are dependable and easy to install. But surface mounted switches on doors and windows are often visible from outside the house. A burglar could try to defeat a magnetic switch by holding his own magnet next to the switch while opening the door or window.

Flush mounted or recessed magnetic switches will eliminate this problem. They are installed into the door, door frame, window frame or sash to hide both the switch and magnet from view.

Recessed Plunger Switches - As an alternative to magnetic switches, plunger switches are sometimes used to protect doors and windows. They are mounted into the door frame on the same side as the door hinges. For windows, they are mounted in the top and bottom of the window frames.

When a door or window is closed, pressure is placed upon the plunger. As the door or window is opened, a spring pushes the plunger breaking a normally closed circuit or closing a normally open circuit. Depending on the type of switch and circuit, an alarm message is sent to the master control panel notifying it of the intrusion.

RECESSED PLUNGER SWITCH
Door Installation

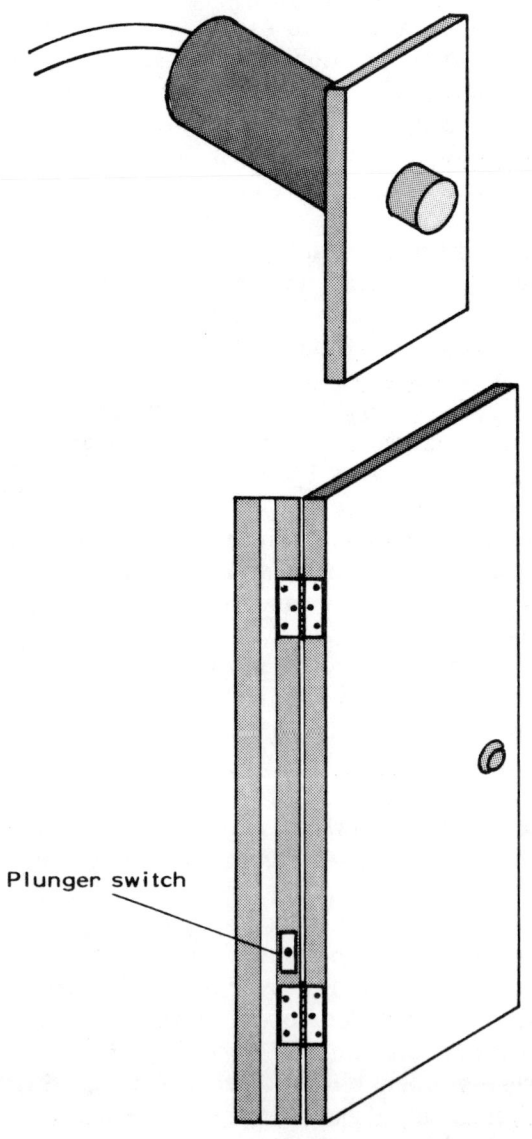

Plunger switch

Plunger switches have the same advantages of concealment as flush mounted magnetic switches, but there are disadvantages. They have more moving parts than magnetic switches making them more likely to fail. Also, the plungers can jam with use and the electrical contacts will corrode over time.

Glass Breakage Sensors - Magnetic and plunger switches are fine for detecting an intruder opening a door or window. But what if the burglar or intruder breaks and removes the glass from the frame and climbs through the opening never activating the switches? Hopefully, the interior sensors will detect him, but we want to stop the intruder before he is able to get inside the home.

When struck, glass will vibrate before it breaks. A glass breakage sensor is able to sense these vibrations and when they reach a predetermined level, the breakage sensor responds by sending an alarm message to the master control panel.

Some of the more sophisticated breakage sensors on the market today use a discriminating circuit to eliminate false alarms because of shocks and vibrations caused by low flying aircraft, sonic booms and heavy equipment. These sensors only respond to breaking glass.

Glass breakage sensors are popular with do-it-yourselfers and are relatively inexpensive and easy to install.

Shock Sensors - These sensors are sometimes used to protect windows and window glass as an alternative to installing glass breakage sensors. Shock sensors are mounted on the inside window frames and react when glass is broken by high energy impact.

Depending on the types of windows in your home, it is sometimes easier and cheaper to buy and install shock sensors. This is especially true if your windows have many separate panes of glass.

Some brands of shock sensors need a signal processor to interpret and relay the alarm messages to the master control panel.

Window Foil - You probably have seen these thin metallic strips bordering the windows of many businesses. Foil has

GLASS BREAKAGE SENSORS

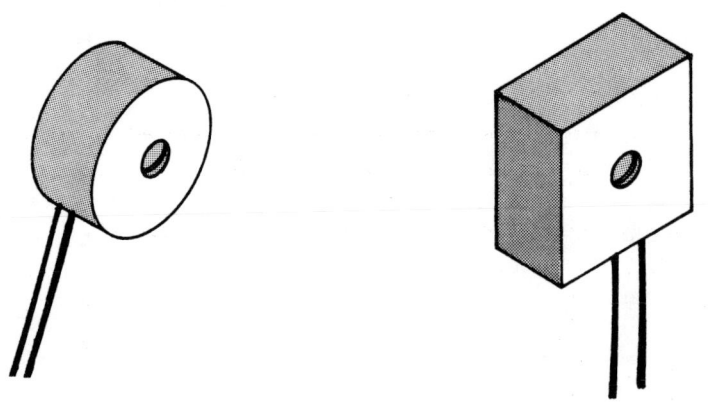

SHOCK SENSORS

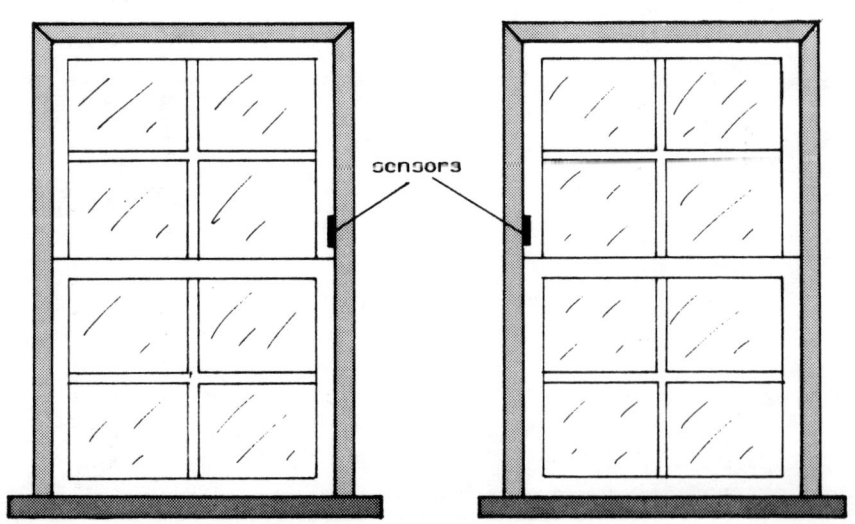

WINDOW FOIL
(view from inside)

been used for years to protect glass windows from burglars and intruders.

Window foil is applied to the glass on the inside and then connected to a normally closed circuit. If the glass is broken or cracked, and causes a break in the foil, the circuit is opened and the alarm message is relayed.

Some professional installers still recommend foil, but for the do-it-yourselfer it has many disadvantages. First, foil takes a great deal of time and patience to apply. It is also easily damaged especially in areas where people and pets come in direct contact with it. In addition, window foil is highly visible in your home. A skilled burglar might try to use a glass cutter to cut around the foil and bypass the alarm system.

A do-it-yourselfer might be better off using glass breakage sensors or shock sensors with magnetic or plunger switches to protect doors and windows.

Wired Window Screens - In addition to installing magnetic or plunger switches on windows, some alarm systems use specially wired window screens. These screens act as a protection sensor as well as keeping insects out of your home.

Wire is woven into a fiberglass screen and then connected to a normally closed alarm circuit. In addition, a magnetic switch is installed at the bottom of the screen and then connected to the alarm circuit. If a burglar or intruder tries to cut or remove the screen, an alarm will be initiated.

These wired window screens have the advantage of letting you open windows while the alarm system is activated providing you can turn off or shunt the window switches. This is especially nice on warm nights, because the screens allow you to leave windows open while still having the protection of your alarm system.

It is not recommended that wired window screens be your only perimeter protection sensors on windows. Use magnetic or plunger switches and glass breakage sensors or shock sensors as the primary detectors and wired window screens as the secondary detectors.

Wired window screens will not completely protect open windows. A burglar with enough time could find and circumvent a wired window screen. When you do go away, even for a short time, close the windows and activate the entire alarm system.

WIRED WINDOW SCREEN
(view from outside)

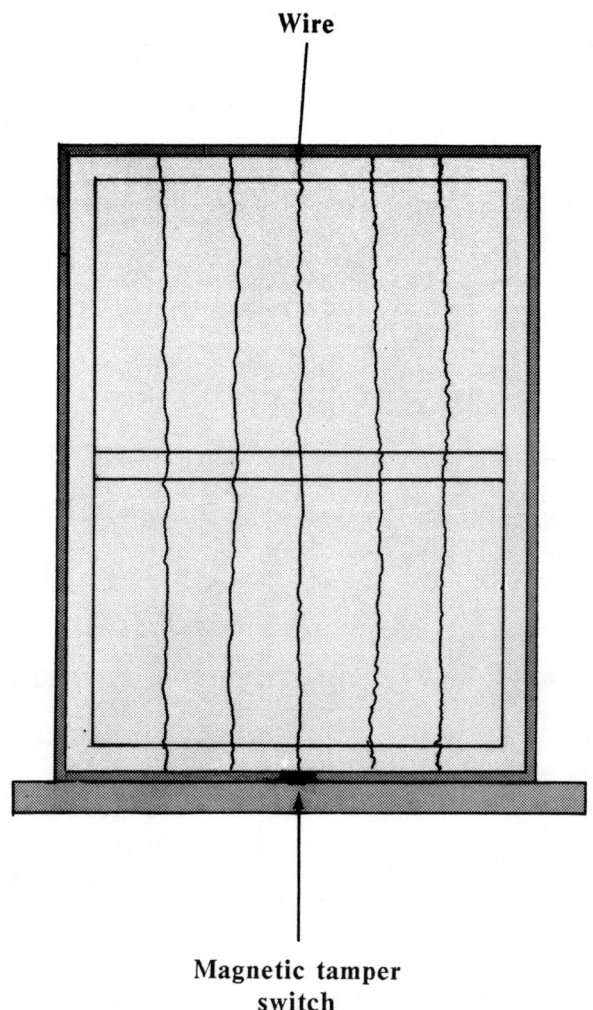

Home Security Systems

Sound Discriminators & Audio Sensors - Sound discriminators, sometimes called audio detectors, are microphones which are sensitive to certain sounds and frequencies. They are tuned to pick out or discriminate for sounds of breaking glass and splintering wood while ignoring other sounds. Sound discriminators are effective because most burglaries involve these noises as the intruder tries to get into your home.

After you adjust the microphone or sensor, the discriminator will measure and compare shock waves and volume to frequency and dispersion. In other words, when the sensor hears the sounds of a burglar breaking in, an alarm message is sent.

The main disadvantage of the sound discriminator is that a burglar or intruder has to break glass or splinter wood to activate a sensor. A skilled burglar might be able to beat the discriminator by quietly forcing a door or window. A professional burglar could pick a door lock without making a sound.

But sound discriminators also have advantages. They can be used to protect a bank of windows eliminating the need to install a glass breakage sensor in every window pane. In addition, a do-it-yourselfer can save time and money by installing a sound discriminator system.

SENSORS & DETECTORS FOR AREA PROTECTION

Microwave Detectors - Microwave detectors use high frequency radio waves to detect intrusion. A transceiver sends and receives radio waves while the detector monitors the reflected energy. An alarm is initiated when the waves sent out have been distorted by someone or something moving in the protected area.

Microwaves have an advantage over ultrasonic detectors because microwaves are not affected by air currents, ringing telephones or small pets. Being sensitive only to bulk motion, such as a person, microwave detectors keep false alarms to a minimum. Microwave detectors are more sensitive when an intruder walks toward and away from the detector, than when he walks parallel to it.

SOUND DISCRIMINATOR SYSTEM

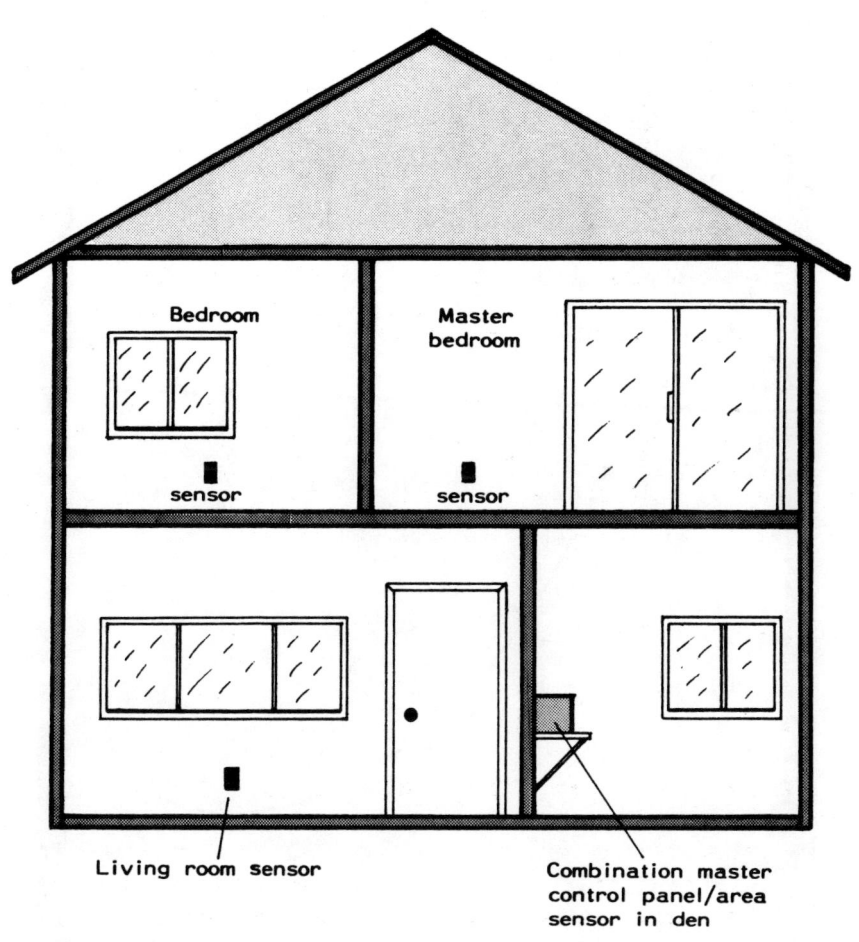

MICROWAVE DETECTOR PATTERN
(view from above)

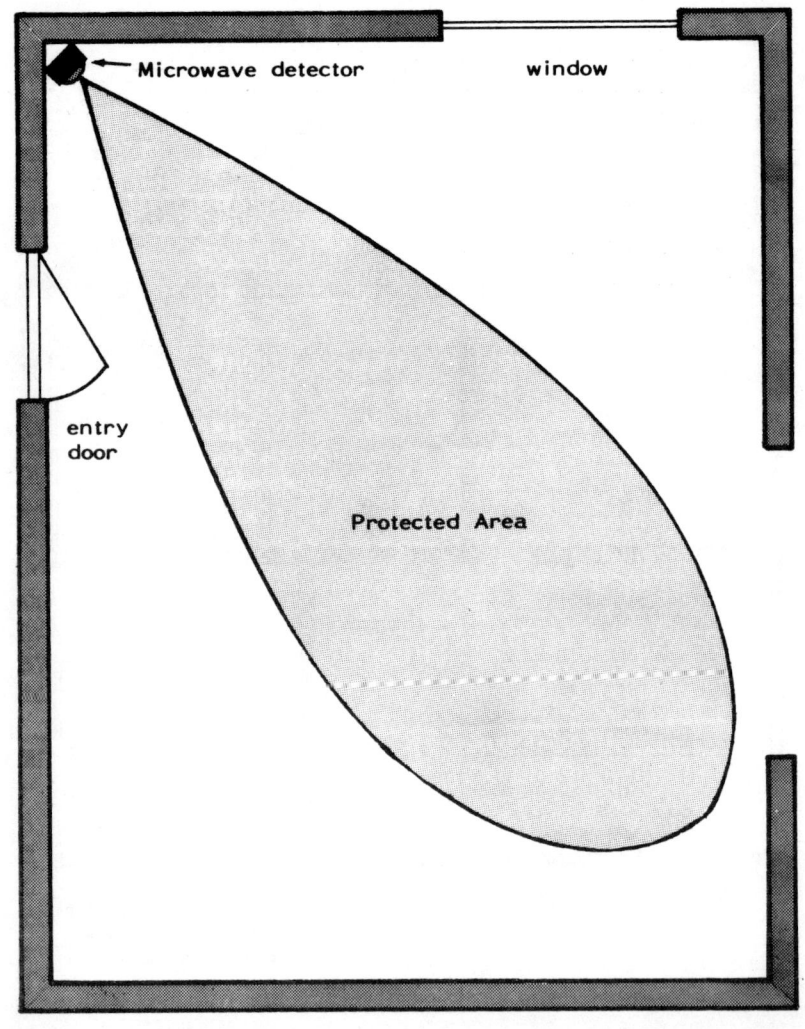

Alarm Components & How They Work

The radio waves transmitted will go through walls and windows. Positioning and adjusting the sensor is very important in preventing false alarms. Someone on the outside of your home could trigger the system if the sensor is pointed towards a window or if the microwaves are allowed to penetrate outside walls.

This penetrating ability can be put to your advantage. Microwave detectors can be hidden behind some solid objects such as room partitions and cabinet doors while still being able to send and receive radio waves. Microwaves will go through wood and glass, but they will not penetrate metal or some plaster walls. Part of the detectors's range will be lost, but it is possible to protect an area from another room or space. Finding and defeating a hidden detector is extremely difficult even for a skilled burglar.

Hidden detectors also keep inquisitive visitors from asking about your alarm system. If they must know, tell them you have a security system, but do not tell them how it works. Who knows who they may talk to. The least number of people who know about your alarm system the better.

One final point about microwave detectors; fluorescent lights give microwave detectors a false signal of movement in the protected area. Either change your lighting or do not use microwaves in rooms with fluorescent lights.

Ultrasonic Detectors - These detectors sense intrusion with sound waves by transmitting and receiving inaudible sound wave patterns. When these patterns are transmitted, they bounce off of ceilings and walls finding their way back to the receiver. The detector compares the sound wave patterns transmitted to those received. An alarm is initiated when the patterns differ. Anyone entering the protected area will cause a change in the sound wave patterns.

The ultrasonic detector is extremely hard to defeat because of its sensitivity to variations in sound waves. This sensitivity is also its main drawback. Small animals in the protected space can set off the alarm system. Air currents from air conditioning and heater vents can also trigger it. Even a ringing telephone can cause the alarm to go off, as well as mail dropping through the door slot.

However, most of these problems can be overcome with proper placement and adjustment. If you install more than one

Home Security Systems

ultrasonic detector, you may need units which can be adjusted to operate on different frequencies. Two sound wave patterns in the same area might overlap and cause one sensor to set off the other.

An ultrasonic detector can be affected by extremes in temperature and humidity. They have a useful temperature range between 50 degrees and 90 degrees Fahrenheit. Before purchasing a unit, make sure the areas you want to protect are not subject to these conditions.

Also, be careful about your pets. Some ultrasonics might disturb them because of the high frequency sound waves that are used to detect intrusion.

Passive Infrared Detectors - These detectors measure rapid temperature changes within a protected area. The sensor does this first by establishing a normal temperature for the area and then monitors it for slight changes. A burglar entering the protected area will give off enough body heat to cause a slight but rapid temperature change. When this occurs, an alarm is initiated. Fortunately, gradual temperature changes will not affect the sensor as the room warms up or cools down during the day.

Unlike ultrasonics or microwave detectors, passive infrared detectors do not send out or radiate energy. They only receive infrared radiation. Multiple infrared detectors working in the same area will not affect each other like some of the other detectors.

Protection is also contained because measured radiation does not penetrate walls, ceilings or other objects. Passive infrared detectors are not affected by air currents, radio waves or noise, helping to minimize false alarms.

Because of these advantages, passive infrared detectors are being used more frequently in home security systems. They are the most sensitive when a burglar travels across the protected area, but care must be taken when installing them. They should not be pointed towards a heat source such as a heater, heater vent, open flame, water heater or an electrical motor which turns on and off by itself. Sunlight or car headlights should not be allowed to hit the sensor directly.

Photoelectric Beam Detectors - Photoelectric beam detectors use a beam of light projected between two points.

PASSIVE INFRARED DETECTOR PATTERN

(view from top)

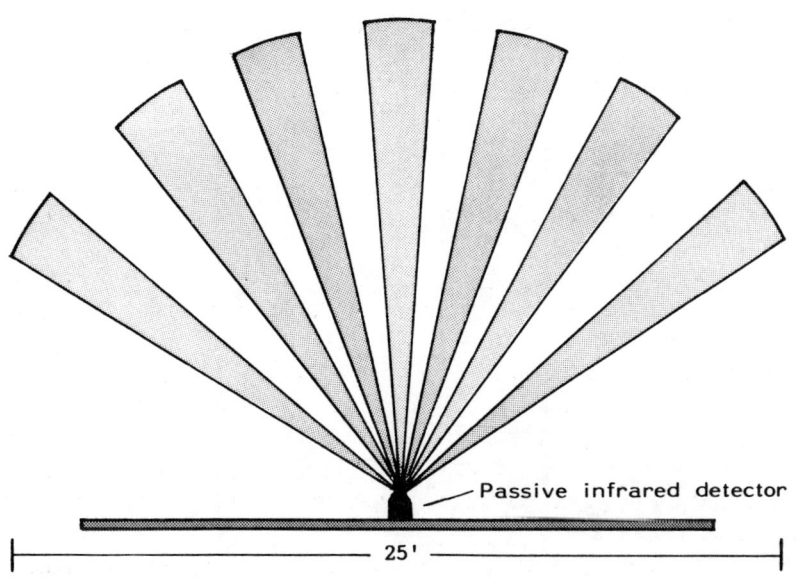

(view from side)

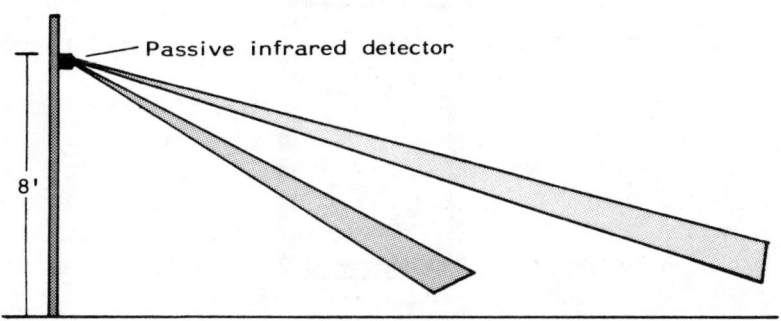

PHOTOELECTRIC BEAM DETECTOR
(hallway installation)

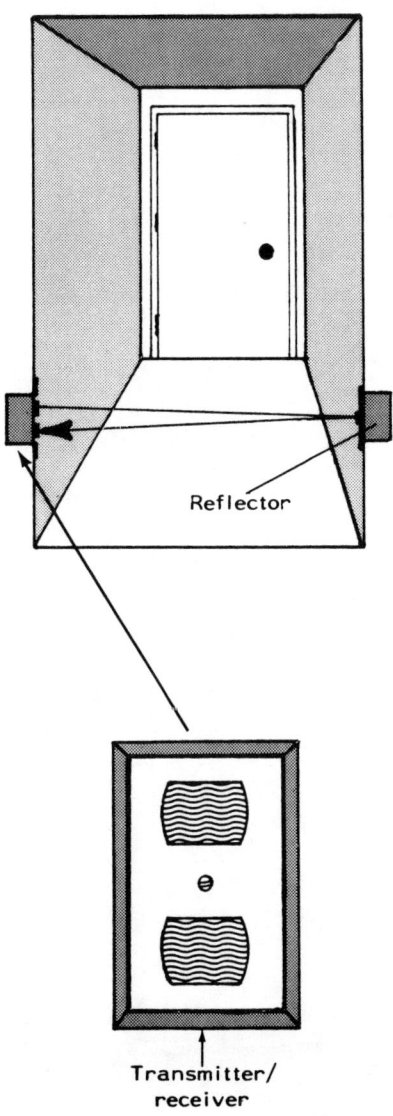

Alarm Components & How They Work

Any interruption of the beam, such as an intruder passing through, sets off the alarm.

Beam sensors are used to protect hallways and stairs. These are areas where a burglar or intruder is likely to pass through in your home. The detector can also be used for perimeter protection by projecting the light beam in front of a row of windows or doors.

A transmitter projects the beam while a receiver monitors the beam. Some units incorporate the transmitter and receiver into a single unit. This single unit uses a reflector to bounce the light beam back to itself. Installation of this combination unit is easier because it eliminates the need to run wires between the transmitter and receiver.

Filters convert the light beam to the infrared spectrum making it invisible to the human eye. Some sensors are disguised as wall outlets so they will blend into a room or hallway. The more sophisticated models use a pulsating beam so the detector cannot be defeated by simply shining another light source into the receiver.

Proper placement is important because if an intruder or burglar suspects a photoelectric beam detector, he might find it before breaking the beam.

Floor Mat Switches - Mat switches are weight sensitive devices that are usually installed under carpets. They are a series of thin metal strips separated by a non-conductive material. When enough pressure is placed on any one of the strips, the alarm circuit is closed and a message is sent.

For area protection, install mats in heavy traffic paths such as in hallways and stairways. Place them in any room or space where a burglar would most likely pass through if he was to get inside your home. Mat switches can be an inexpensive means for protecting an area, room or space.

Mat switches come in standard widths to fit in most hallways. You buy them by the foot allowing you to install just about any length you need. The entire series of switches is enclosed in plastic which helps when installing them under carpets. The plastic also keeps moisture away from the contact points while increasing the mat's useful life.

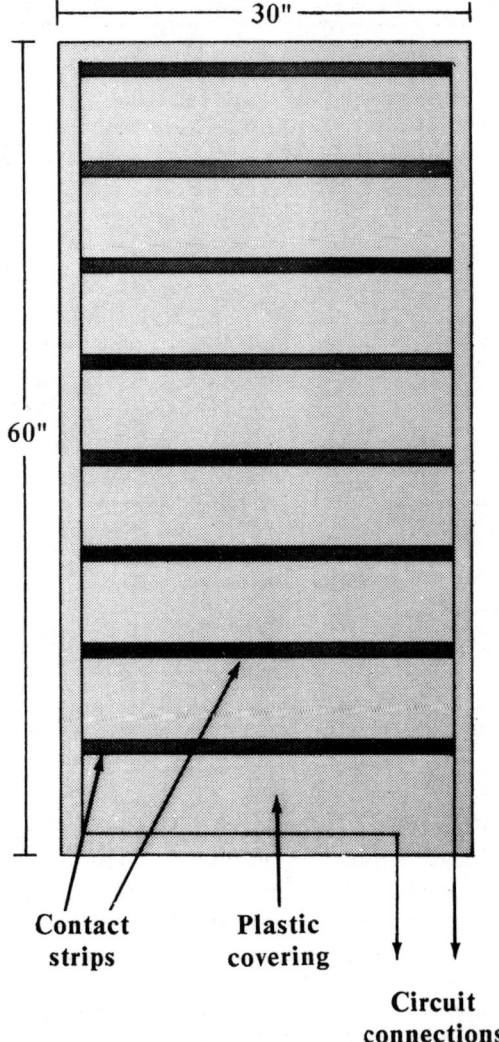

COMPARING AREA SENSORS

Conditions affecting the protected area	Microwave	Ultrasonic	Passive infrared	Photoelec. beam
Warm to hot temperatures	no effect	some effect	range is reduced	no effect
Excessive humidity	no effect	some range reduction	range can be affected	no effect
Vibrations	can be a problem	some problems	problem if excessive	problem if excessive
Noise	no effect	can be a problem	no effect	no effect
Air currents and drafts	no effect	placement important	placement important	no effect
Heating and air conditioning vents	no effect	placement important	placement important	no effect
Stray light hitting the sensor	no effect	no effect	can be a problem	can be a problem
Moving draperies or fan blades	placement important	placement important	little or no effect	placement important
Radio frequency interference	can be a problem	can be a problem	can be a problem	can be a problem
Penetration of walls and windows	can be a problem	no effect	no effect	no effect
Metalic reflection (foil wallpaper)	can be a problem	little if any effect	little if any effect	little if any effect
Interference between sensors	can be a problem	can be a problem	no effect	no effect
Sensitivity to pets	can be a problem	can be a problem	adjust above pet	install above pet
Adjustable range	yes	yes	no	no
Power consumption (for batteries)	medium to high	medium	low	medium
Cost relative to other sensors	average	average	average to high	low to average

This table should only be used as a guide for comparing sensors. Your home might have other conditions which affect their performance. Consult the manufacturer's specifications for your exact requirements.

Home Security Systems

Mat switches use normally open circuits but can be converted to normally closed circuits to interface with other protection components, circuits or zones.

SPECIAL PROTECTION SWITCHES & SENSORS

Panic Switches - Panic or emergency switches are used to sound alarms instantly. This can give you immediate protection if an intruder tries to break in while you are at home.

You may want a panic switch near your front door and near your bed. At your front door a panic switch is important, because if a stranger tries to force his way past you, the alarm can be sounded. At night if you hear a burglar in the house, a siren or bell can scare him away. You can also use the panic switch to turn on inside and outside lights.

Hand-held transmitters can also be used to initiate an alarm. Because of their portability, they offer greater flexibility in your personal safety.

Smoke & Heat Detectors - If you are planning a home security system, using smoke and heat detectors is a wise decision. They make your alarm system more versatile. For the extra protection gained by their installation, your money is well spent.

Even if you do not plan to install a security system your home should be equipped with them. These detectors do not prevent or put out fires, rather, their purpose is to warn the family to get out of a burning house before it is too late.

This early warning is especially important at night because you could be overcome by smoke while sleeping and not get out of your home safely. Many times smoke is a bigger threat to life than fire.

Your alarm system should include detectors located throughout your home. Each bedroom should have its own detector, and each level of a multi-level house also should have at least one detector. Other smoke detectors should also be installed between the sleeping areas and locations where fires are likely to start.

Smoke detectors are available that are self-contained alarms. That is, they have their own power source and alarm horn built into the unit. Other detectors are designed to

Alarm Components & How They Work

interface with a master control panel's fire protection circuits. (Remember, not all master control panels have fire protection circuits so plan your system accordingly.)

More fire protection is provided when your home is monitored by a central station. The fire department can be summoned quickly especially when no one is home. Quick response can prevent extensive damage to your home and possessions.

When properly located and installed, smoke detectors provide a warning before a fire presents a life threatening situation. Two types of smoke detectors provide this protection. Each senses different types of fires and smoke more efficiently. They are the photoelectric and ionization smoke detectors.

Photoelectric Smoke Detectors - These detectors sense smoldering fires better than the ionization detectors. Photoelectric detectors work by sending a light beam into a sensing chamber. As smoke enters this chamber, light is reflected off of the smoke particles. The detector reacts and initiates an alarm when enough particles are present to reflect a predetermined amount of light.

Photoelectric detectors react better to smoldering fires because this type of fire produces larger smoke particles which reflect more light. The ionization detectors work by a different method.

Ionization Smoke Detectors - Fast burning fires are detected more quickly by these smoke detectors. They react better than the photoelectric detectors to smaller, less visible smoke particles from rapid burning fires.

Ionization detectors use a very small amount of radioactive material to increase the electrical conductivity of the air in the detector's chamber. As smoke enters this chamber, the electrical current flowing through the chamber is reduced. A fire alarm is initiated when the reduction drops below a predetermined level.

Heat Detectors - There are two different types of heat detectors available. The first is the rate-of-rise detector which senses a specific temperature increase over a fixed time period. When the detector senses this change, contacts are closed and the alarm is initiated. The second type of heat detector reacts at a specific temperature: usually around 135 degrees Fahrenheit. At this temperature a metal strip or coil expands or changes shape, closing the contacts and activating the fire alarm circuit.

Install heat detectors in areas where smoke detectors will not work properly or will be subject to false alarms. For example, install them in kitchens where cooking smoke could set off a smoke detector. They can also be used in furnace rooms, attics and garages where smoke and dust would cause false alarms.

Heat detectors should be used with smoke detectors and not instead of smoke detectors. Remember, a heat detector only senses heat. You could be overcome by smoke long before a heat detector would react.

TAMPER SWITCHES

Tamper switches are used to detect any attempt by a burglar to remove or open enclosures protecting alarm components, such as: control panel doors, switch plates, junction box covers, and alarm device housings. Also, they are used to protect wired window screens. These switches initiate an alarm if an intruder tries to defeat the alarm system before trying to get inside the house.

ALARMS & ALARM TRANSMITTING DEVICES

Local Alarms As stated before, local alarms are used to scare away the burglar or intruder. Many alarm systems use this audible alarm with an automatic telephone dialer or digital communicator. Without the local alarm you would only have a silent alarm response. This is fine when we want to catch the burglar in the act, but our primary purpose is to keep the intruder out of the house and prevent a dangerous confrontation.

Electronic sirens are used as local alarms which imitate the sounds of an emergency vehicle, such as a police car siren. With these sirens the frequency changes to make the alarm

Alarm Components & How They Work

more distinct. This sound change is referred to as a "yelp" or "rise and fall".

Bells are also used as local alarms. They can be connected to the fire protection circuit to warn your family and the neighbors, as well as distinguishing the emergency as a fire rather than an intrusion.

Any local alarm you install must be very loud. Inside the house you want to make it painful to the burglar's ears. Outside you want to be sure one of your neighbors hears the alarm and calls the police.

Flashing strobe lights are sometimes installed in addition to sirens and bells. The flashing lights attract attention and makes it easier for the police to spot which house the alarm is coming from. With only a siren or bell it is sometimes difficult to determine just which house needs help.

Automatic Telephone Dialers - These alarm devices are sometimes called tape dialers. They send a prerecorded tape message over regular telephone lines to warn that an emergency is taking place and to identify the home and its location.

Some dialers have two input channels and two different taped messages. One input is connected to the fire protection circuit and the other is connected to the burglary or intrusion circuit. The dialer with this dual system can then request either police or fire response.

Equip your dialer with a battery backup unit to insure that the system will not be compromised in the event of a power failure.

Other features and options found on automatic telephone dialers include the ability to:

1) Seize the telephone line and take priority over other calls.
2) Dial more than one telephone number for each output message to insure that the taped message is delivered.
3) Redial a busy telephone number when one is encountered.
4) Reset the dialer automatically after delivering an alarm message.

Home Security Systems

5) Abort the alarm message if the homeowner happens to accidentally trip the alarm system.

Some dialers can be used to listen in on a home after the alarm is tripped and the dialer relays the taped message. The telephone line is left open, allowing the person receiving the call to hear if a burglar is in the house.

Digital Communicators - Digital communicators differ from telephone dialers in that they communicate with a central station by electrical impulses. The communicator first captures or seizes the line like the dialer and then dials the central station to transmit the alarm message. The central station monitor receives the alarm information on a receiving panel and logs the alarm as to the homeowner, his address, type of emergency and the time the message was received. Once the monitor confirms that the alarm is not false or accidental, he calls for police, fire or medical response.

Some communicators have an abort time delay. This grace period between detection and transmission allows the homeowner time to shut down or reset the system if it was accidentally tripped. In addition, some communicators have a listen-in feature similar to the automatic telephone dialer.

Telephone line seizure is very important for both the digital communicator and the automatic telephone dialer. It prevents a burglar from defeating an alarm system from transmitting its messages by taking a telephone off the hook, or by calling your number to tie up the line.

Telephone lines used for digital communicators and automatic telephone dialers can also be monitored for service disruptions. With a supervised telephone line, the central station would know there was trouble with the connection but would not know if the line was cut by a burglar, or if it was simply out of service. Local alarms could be sounded if the telephone line was cut and the central station could dispatch someone to investigate.

There are other methods for relaying alarm messages, but most homeowners will use a digital communicator or an automatic telephone dialer with local alarms. Some won't bother with telephone dialers or digital communicators; they feel a local alarm is adequate for their security needs.

Chapter 8

Planning Your Security System

At this point you should have a basic understanding of perimeter and area protection systems. You have seen how alarm messages are relayed to the master control panel by wires, wireless transmitters and line carriers. Local alarms, automatic telephone dialers and digital communicators were also described showing you how alarm messages are transmitted and how help is summoned.

Before installing any type of security system, professional security companies will conduct a threat analysis. They want to know what the alarm system is to protect and from whom. For a home security system you should have an idea who and what will threaten your family and possessions. (Most home owners are concerned about burglars and vandals as well as fires.)

Also, you should consider what you want your alarm system to do. For example, do you want it to simply scare off intruders or do you want it to summon the police, the fire department and medical help as well?

DETERMINING YOUR SECURITY NEEDS

In planning your alarm system you first need to know how to determine your security needs. Every home and homeowner is different. The type of security system you install could differ substantially from your neighbor's system, from the components to the level of response.

Home Security Systems

The first thing you need to do is to list of all your possessions. Do you have expensive jewelry, antiques, cash, silver or gold in your home? It may surprise you to learn just how much you could lose. Burglars have been known to completely empty a house while the homeowner was on vacation. Everything in your home could be taken including: furniture, appliances and clothing. The more valuables you have, the greater your chances are of being burglarized. Combine this with a house that is not protected with strong door and window locks and a burglar alarm system and your home becomes a prime target for burglars. The more you have, the more security you will need.

More important than possessions is the safety of you and your family. A confrontation with a burglar in your home could be very dangerous. Deterring a burglar with an alarm system can prevent a confrontation in the first place.

A security system minimizes your chances of walking in on a burglar, but it can also protect you when you are in the house. By activating only the perimeter switches and sensors and by shutting off the area detectors, you can move about your home in safety. If an intruder tries to break in, the perimeter system will initiate an alarm. Hopefully, this will scare him away and give you time to call police. Personal safety deserves much consideration when determining your security needs.

Peace of mind is also very important. Do you worry about someone breaking into your home while you sleep? When you are away from home, at work or on vacation, do you worry about your home being burglarized? Installing even a simple security system could set your mind at ease.

The neighborhood you live in also determines how much of a security system you need. Obviously, if you live in an urban area with a high crime rate, you will need more security than if you live in a rural area with little crime. Check with your police department to find out what the crime rate is for your community. The higher the crime rate, the more security you need.

Who is watching your home? Potential burglars are, but hopefully your neighbors are keeping a watchful eye on your home when you are away. If you or a member of your family is usually home at all times, your security needs will be less than if your home is frequently left unattended.

Planning Your Security System

If someone is usually home and if you know your neighbors and can depend on them to watch your home, a local alarm might be adequate for your needs. If you are gone frequently and do not have someone to watch over your home you might consider central station monitoring.

Before we get to the planning of an alarm system we first have to look more closely at the different levels of security and protection.

LEVELS OF PROTECTION
Different Alarm Response

Local Alarms - This is the cheapest alarm to install and maintain, but by itself provides only minimum security. Located inside and outside your home, alarm sirens and bells scare burglars and intruders away and signal neighbors to call police. Many times these local alarms are adequate for the average homeowner and a large number of do-it-yourselfers install them as their only alarm device.

Local alarms deter burglars by drawing attention to the intrusion. If you install only a local alarm be sure to instruct the neighbors to call police if they hear the siren or bell. It is good practice anyway to know your neighbors and to watch each other's homes when either of you is away.

Local alarms are used with other alarm devices such as automatic telephone dialers and digital communicators. Local alarms alone could provide you and your family with enough security, but what if your home is quite a distance from your nearest neighbor and that neighbor could not hear the siren or bell?

Automatic Telephone Dialers - The dialer might be the answer if your nearest neighbor is too far from your home to hear a local alarm. When an intrusion is sensed, the dialer is programmed to dial out over the regular telephone lines. The dialer seizes the line and takes priority over other calls. When a connection is made the dialer plays a prerecorded taped message giving your name, address and a message that emergency help is needed.

Some automatic telephone dialers can request either the police or fire department response. They will dial more than

Home Security Systems

one telephone number making sure that someone will receive the message. Also, the dialer keeps dialing the number if it gets a busy signal.

You could program the dialer to call you at your office, a neighbor next door, the police or the fire department. However, most police departments do not respond to taped messages because of the high incidents of false alarms. Others assign a low priority for response. If you install an automatic telephone dialer, contact the police department and fire department before programming in their telephone numbers. Some cities forbid dialers placing calls directly to the police and fire departments. Also you should check with the telephone company for any installation requirements or restrictions.

Automatic telephone dialers provide more security than local alarms alone, but someone must be available to receive the taped message and then call for the police or fire department response.

Central Station Monitoring - A central station can provide you with even more security. Your home is watched 24 hours a day from a remote location. The central station receives alarm messages and information from your digital communicator. The communicator first identifies you and your address and then tells the monitor what type of alarm condition exists: intrusion, fire or medical. The station monitor will then call the police or the fire department after waiting a few moments for an abort signal or cancel call. The monitor will then try to call you at work informing you of the alarm. If you cannot be reached, he can be instructed to call a neighbor or relative.

Some central stations have a call-in service for vacationers and business travelers. You can inquire about your home from anywhere in the world.

In addition, the central station can check on your alarm system to see that it is working properly. The station monitor can tell if your standby batteries are low or if the protection circuits and zones are functioning normally. (Your alarm system has to have the proper equipment and features to perform this function however.)

When the monitor calls and tells police of the alarm condition, you hope they will rush to investigate. But how long

Planning Your Security System

will it take for the police to respond? Some of the more expensive central station monitoring services have their own private security patrols. Dispatched by radio, these armed guards can sometimes get to your home faster than police.

Whether you subscribe for private armed guard response or if you are going to rely on the police, it is important to find out just how long it will take for help to arrive once an alarm message is sent. (This may shock you when you see how long it can take!)

SUMMARY OF SECURITY NEEDS & RESPONSES

It is very important to get the right protection for your individual security needs. Recommending one type of alarm system would be impossible without a complete study of your home and life style. Every home is different and every homeowner's requirements vary. You have to take into consideration what you have to protect, how much you can afford to spend, where you live and your life style.

In planning a home security system, you have to determine the following:

What Protection System(s) to Install
- Perimeter Protection
- Area Protection
- Combination of Perimeter/Area Protection
- Special Protection: Fire, Panic, Medical

What Alarm Response You Need
- Local Alarms only
- Automatic Telephone Dialer with Local Alarm
- Central Station Monitor via Digital Communicator with a Local Alarm

Method of Communication to the Central Station
- Regular Telephone Lines
- Radio Transmitter
- Two-way Telephone Cable
- Dedicated Telephone Lines

Who Will Respond to an Emergency
Yourself
Your Neighbors
The Police
The Fire Department
Paramedics
Private Armed Guard

As you decide about the protection systems, response, method of communication and who will respond, you have to consider:

1) Your family's safety
2) Your possessions
3) Where you live
4) Who watches your home when you are away

PLANNING AN ALARM SYSTEM FOR YOUR HOME

Here are the steps you use to plan and install a home security system. The following example is a combination perimeter and area system with fire protection sensors, emergency panic switches and local alarms. *Remember, this is only an example and may not be adequate for your security needs.* Other sensors and alarm devices may be needed.

Floor Plans - The first step is to draw floor plans of your home. Include second stories, basements and garages. This floor plan should be drawn to scale showing the location of all doors, windows and skylights, any place a burglar or intruder could enter. Draw all rooms, hallways and stairs and show the locations of all walls and room partitions.

You will use the floor plan to determine:

1) The number of switches and sensors for a perimeter system
2) The number and location for area sensors
3) The number and location for fire detectors and emergency panic switches
4) The location for the master control panel

Planning Your Security System

5) The locations for the local alarms
6) The locations for the access panels
7) The locations for running wires between the master control panel, access panels switches, sensors and alarm devices
8) The locations for other alarm equipment you may choose to install but not shown in this example. Other alarm equipment you might choose to install could include an automatic telephone dialer, digital communicator and wireless transmitters.

Perimeter Protection - The second step is to decide what type of switches and sensors to install for a perimeter system. You have these basic choices:

Doors - Magnetic switches, plunger switches

Windows - Magnetic switches, plunger switches, glass breakage sensors, wired window screens, sound discriminators, window foil

Now refer back to your floor plan. Starting with the front door make a list of all the doors, windows and openings that are on outside walls. For the windows make a note as to the type and size. You need to know the window configurations to estimate the total number of glass breakage sensors, the size of wired window screens and the amount of window foil you will need.

Beside each door and window on your list write down the type of switch or sensor you will need to protect that entry point. This list now becomes your shopping list for the perimeter switches and sensors.

Area Protection - Next, study your floor plan to determine where a burglar would pass through while looking for valuables. Hallways, stairways and the master bedroom are areas a burglar would probably go through in your home. The purpose of all this is to determine how many area sensors you will need and where to locate them. Make a room by room list like the one made for the perimeter switches and sensors.

PERIMETER PROTECTION SAMPLE FLOOR PLAN
Locating Switches & Sensors

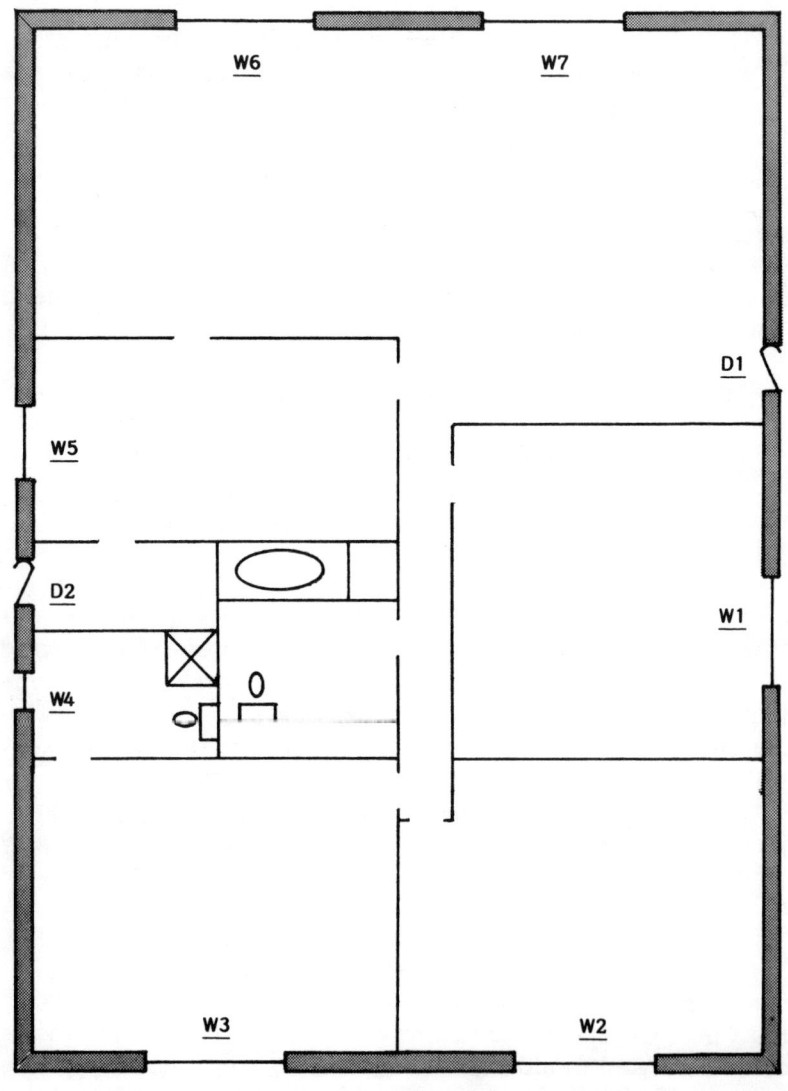

KEY TO PERIMETER PROTECTION SENSORS

Room	Point of Entry	Comments
Living room	Door D1	Surface or recessed magnetic, or plunger switch
	Window W6	Non-opening aluminum frame 48"x60"
	Window W7	Glass breakage or shock sensors, or sound discriminator
Front Bedroom	Window W1	Sliding aluminum frame 30"x60" Surface mounted magnetic switches, glass breakage or shock sensors, or sound discriminator Wired window screens
Back Bedroom	Window W2	Sliding aluminum frame 30"x72" See Window W1 for switch and sensor choices
Master Bedroom	Window W3	Same as Window W2
Master Bath	Window W4	Sliding aluminum frame 30"x36" See Window W1 for switch and sensor choices
Kitchen	Window W5	Sliding aluminum frame 30"x48" See Window W1 for switch and sensor choices
Service Room	Door D2	See Door D1 for switch choices

AREA PROTECTION SAMPLE FLOOR PLAN
Locating Sensors, Switches, Alarms & Panels

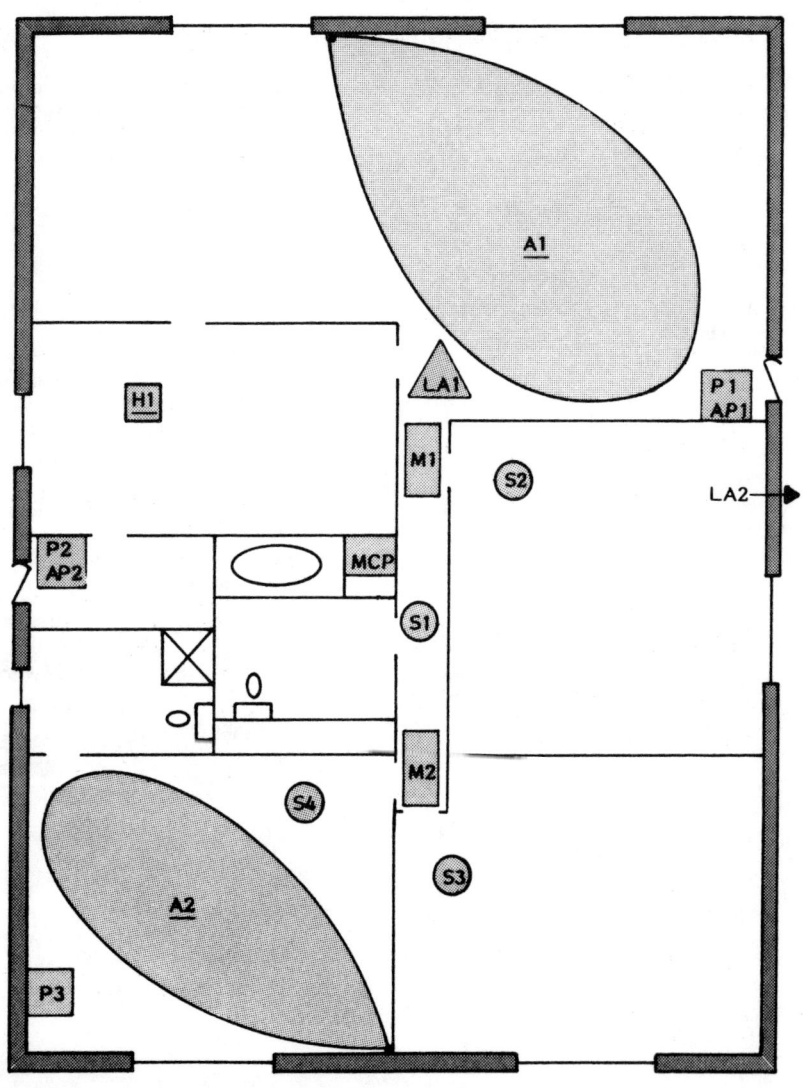

KEY TO AREA PROTECTION SENSORS

	Location Symbol	Comments
Area Sensors		Choices: microwave, ultrasonic, passive infrared, photoelectric beam
Living room	A1	
Master bedroom	A2	
Mat Switches		Determine sizes needed
Main hallway	M1	
	M2	
Smoke Detectors		Select types: photoelectric, ionization
Main hallway	S1	
Front bedroom	S2	
Back bedroom	S3	
Master bedroom	S4	
Heat Detectors		Select type: specific temperature, rise over time
Kitchen	H1	
Panic Switches		Determine where to mount, e.g. combined with access panel, wall switch, portable
Front door	P1	
Back door	P2	
Master bedroom	P3	
Master Control Panel		Determine zones, circuits, features, connections, etc.
Hall closet	MCP	
Access Panels		Choices: inside or outside, key or key pad
Front door	AP1	
Back door	AP2	
Local Alarms		Choices: sirens, bells, separate fire alarm
Inside alarm	LA1	
Outside alarm	LA2	

Home Security Systems

You should decide on using one or more of the following detectors to protect the interior:

1) Microwave
2) Ultrasonic
3) Photoelectric Beam
4) Passive Infrared
5) Mat Switches

Remember, you have to consider the characteristics of each area or room you want to protect. The area sensors listed above will produce false alarms if they are incorrectly chosen or improperly installed. (Refer back to Chapter 7 to determine which sensors meet your requirements.)

Emergency Panic Switches - If you decide to install panic switches, you might want to locate one near the front door. Position it so you can reach it while opening the door. In the master bedroom your panic switch should be near your bed and within reach so you do not have to get up to activate the alarm or turn on the lights. You can also install another panic switch in the central part of your home. Refer back to your floor plans and mark these location.

Smoke & Heat Detectors - For maximum, safety smoke detectors should be installed in every bedroom as well as in other strategic locations throughout your house. You want to be warned as soon as a fire starts. You can choose between photoelectric and ionization detectors to protect your home and family.

Consider installing heat detectors in areas like the garage, attic and furnace room; anyplace a fire could start and where a smoke detector would be subject to false alarms.

Locating the Master Control Panel, the Automatic Telephone Dialer and the Digital Communicator - The next step is to determine the location for the master control panel. You want to install the panel in an out-of-the way place. Again, refer to your floor plan. Try to find a central location which is hidden and out of normal traffic patterns. Under a stairway or in a secluded closet are good locations. Remember, you are going to run wires from the sensors to the panel so

Planning Your Security System

find a location that is near the center of the house to reduce the length of wiring runs.

You will also want a central location if you use wireless transmitters. This minimizes the distance that the radio waves must travel.

If you are going to install an automatic telephone dialer or digital communicator, you will want to install them near or next to the master control panel. It is a good idea to keep these components hidden and out of sight as well.

The location for the master control panel, automatic dialer and the digital communicator will need a 120 volt power source. (In other words a wall outlet.) Also, a telephone line will be needed for the dialer and the communicator.

Access Panels - After you have determined where to install the master control panel, the automatic telephone dialer and the digital communicator, refer to your floor plan to decide which doors you use most frequently. Install the access panels at these entries to make using your alarm system more convenient and to remind you to arm the system when leaving and to disarm it upon returning.

Mark the locations for the access panels on your floor plan.

Local Alarms - Next, find the location for the inside local alarm. It should be hidden from view and installed away from the master control panel, telephone dialer and the digital communicator. The idea is to keep an intruder from being drawn towards the panel, dialer or communicator when the alarm sounds. Behind a dropped ceiling or in an air return vent are good locations.

The outside local alarm should be on the street side of your home. It is very important that your neighbors and passing police will hear it. Mount the alarm high enough so it cannot be reached by a burglar or install it behind an attic vent.

Wiring - The final step is to use the floor plan to determine where to make the wiring runs. Remember, you will have to connect the switches and sensors to the master control panel and then connect the access panels and the alarm devices.

Home Security Systems

You might be able to run wires through the attic or under the house. Use the floor plan to draw a tentative wiring diagram. If you find that the wiring is going to be difficult, consider using wireless transmitters.

At this point you might think the job of installing a security system is too complicated for you. Maybe you do not have the electrical or mechanical skills. But don't despair, Chapter 10 shows you how to find a professional alarm company to do the planning, installing, monitoring and servicing of your alarm system.

Homeowners with basic skills can install their own alarm systems. While the majority of these do-it-yourself systems have only local alarms, some use automatic telephone dialers. Only a few homeowner installed systems are monitored by digital communicators.

If you need central station monitoring, check with the monitoring services before you buy and install your system. Some central stations refuse to monitor homeowner installed systems, while others require that your alarm system be inspected and tested before providing monitoring services.

If you can find a central station that will monitor a system you installed yourself, you can save money and protect your home with a high level of security.

REMEMBER THESE POINTS

1) Buy good equipment. It costs less in the long run.

2) Use the components for the job they were manufactured to do.

3) Read all instructions supplied with the equipment *before* trying to install it.

4) Install the equipment according to the manufacturer's instructions and recommendations.

5) Make and double check all connections before applying power.

Chapter 9

Do-It-Yourself Installation

By now you should have decided upon perimeter, area, special or a combination of protection systems. You should have a good idea what components you are going to install and what response you will need. In addition, your floor plans and equipment lists should be complete.

PERMITS & LICENSES

The first step before purchasing and installing any alarm equipment is to check with your local building department. You might need a permit for your alarm system installation. Also find out what building and electrical codes you must comply with.

Next, determine if your city, county or state has laws and regulations governing alarm systems. Do this by calling the licensing and permit divisions. You may be required to pay a fee and obtain a license for the alarm system.

SHOPPING FOR ALARM COMPONENTS
Where To Find Them

To install a home security system you will need:

1) A Master Control Panel
2) Switches & Sensors for doors and windows
3) Area & Room Protection Sensors

Home Security Systems

4) Smoke & Heat Detectors
5) Panic Switches
6) Local Alarms
7) Installation Tools
8) Installation Supplies

Optional components you may choose to install:

1) Special Protection Sensors (gas, water, etc.)
2) Wireless Transmitter & Receivers
3) An Automatic Telephone Dialer
4) A Digital Communicator
5) Access Panels & Shunt Switches

One of the best places to start looking for security alarm equipment is in the telephone book yellow pages under: *"Alarms"*, *"Burglar Alarms"* or *"Security Equipment"*.

The best source for alarm equipment is from a store specializing in burglar alarms and security equipment for the do-it-yourselfer. This type of specialty store will have a large selection of brands and components and can give you help if you encounter problems designing or installing your alarm system.

If you do not have a do-it-yourself security store in your area, you can purchase security alarm components and installation supplies by mail. Contact:

Mountain West
4215 N. 16th St.
Phoenix, AZ 85064

Ask for their latest catalog of home security equipment and installation supplies.

Do-It-Yourself Installation

HOW TO CHOOSE ALARM COMPONENTS

Burglar alarm components designed for the do-it-yourselfer are tested and written up in different consumer magazines. Try checking the year end issue of **Consumer Reports Magazine** and their **Buying Guide** for security and burglar alarm equipment evaluations. Also, check magazines like **Popular Science, Popular Mechanics** and **Mechanics Illustrated** for assessments.

Be sure to take your floor plans and equipment lists with you when shopping for alarm components. An experienced salesman might be able to show you how to improve your plans and explain how to save money on the installation. Ask plenty of questions about the different types and brands of equipment. But don't settle for only one salesman's opinion. It pays to shop around.

Look for security components and equipment which are listed by Underwriters Laboratories. Equipment without a UL listing does not mean that it is not good and should not be purchased. It is possible that a UL listing is forthcoming. Try to find out if the equipment is new on the market or if the manufacturer did not bother to apply for the listing. (See Chapter 10 for a discussion of Underwriters Laboratories and the security industry.)

When choosing alarm components, make sure that the sensors, switches, master control panel, access panels, automatic telephone dialer, digital communicator, wireless transmitters, receivers and the local alarm bells and sirens will be compatible. One manufacturer's access panel might not work with another's master control panel. The point is to ask questions and find out about the alarm equipment before you buy and attempt to install it. And don't assume that you can return the alarm equipment after you purchase it.

Also inquire about the different manufacturers' warranties and service policies before purchasing any alarm equipment you are going to install yourself. It is a good idea to know where to get parts and service before you may need them.

Home Security Systems

HOW TO PURCHASE ALARM COMPONENTS

Purchasing the Master Control Panel - The type and size you buy depends on how big your home is and how many switches and sensors you plan to install. (This is why you made a list of all door and window switches and sensors, as well as all area and special protection sensors and detectors.) Each protection circuit has a maximum resistance that should not be exceeded by adding extra switches and sensors to it. First, determine how many and what type of switches, sensors and detectors you are going to install. Then purchase a master control panel which will have enough circuits and zones to accommodate them.

If your master control panel is going to supply any of the sensors with a stabilized power source, do not exceed the control panel's current rating. First, determine the total draw for all the sensors and then buy a panel that meets or exceeds the total power requirements of the entire system.

If you plan to install a zoned system, make sure the control panel will have enough separate alarm circuits for your needs. Also, if you intend to install a telephone dialer or digital communicator, the control panel must have the proper output connections for them. The access panels will have to interface with the master control panel as well.

The point is to determine what sensors, switches, alarms and other equipment you are going to install *before* buying the master control panel. Then purchase a panel which has the circuits, features, connections and power requirements you need.

Look for the following when buying a master control panel:

Circuits:
    Normally open circuits (N.O. or O.C.)
    Normally closed circuits (N.C. or C.C.)
    Delay entry/exit circuits
    Day/night circuits
    Panic circuits (24 hour)
    Fire protection circuits

Features:
    Lockup relays

Do-It-Yourself Installation

    Manual reset switch
    Automatic shut off and reset
    Alarm system status lights (LED's)
    Battery backup power source
    Test switch

Connections:
    Local alarms
    Automatic telephone dialer
    Digital communicator
    Access panels

Some master control panels on the market for the do-it-yourselfer combine an area sensor such as a microwave or ultrasonic into one unit that also functions as the control panel. These combination units help to reduce the cost of an alarm system, but they do not have the versatility of a separate master control panel. Before you buy one of these units, make sure it has all the circuits, features and connections you need. Don't settle for an alarm system that is inexpensive and will not properly protect your family, home or possessions.

Purchasing Switches & Sensors - Don't complicate your system more than necessary by installing different types and brands of sensors. Try to design your system to use similar components. But, if you have a special need for a different switch or sensor, by all means use it.

Area sensors should "fit" into the room or space they are going to protect. For example, microwave sensors should be contained and not allowed to spill into other areas which could produce false alarms. Ultrasonics should not be used in areas where small pets or moving curtains could trigger an alarm. Passive infrared sensors should not be used where sunlight, headlights or temperature changes will create false alarms. First, survey the rooms and spaces you intend to protect with area sensors then look for conditions that will trigger false alarms. Select and purchase areas sensors using the comparison chart in Chapter 7 as a guide.

Purchasing Local Alarms, Automatic Telephone Dialers & Digital Communicators - When purchasing these alarm devices, you have to be sure they will be compatible with your master control panel. The control panel has to have the proper output connections and power requirements for the local alarms, the automatic telephone dialer and the digital communicator. Before buying, compare the manufacturer's specifications for both the control panel and the alarm devices so you will know if the units will work with each other. If the power fails, the standby batteries must also have sufficient capacity to power the digital communicator, the automatic telephone dialer and the local alarms. When purchasing these alarm devices make a note as to their power requirements so you can determine which standby battery to buy.

Look for these features in alarm components:

Local Alarms:
- High noise output as measured in decibels
- Electronic sirens that "yelp"
- Weatherproof construction for outside alarms
- Outside alarm boxes with tamper switches

Automatic Telephone Dialers:
- Line seizure capability
- Redial if line is busy
- Multiple number dialing
- Automatic reset
- Abort capability
- Dual input/output for intrusion & fire
- Listen-in feature
- Silent operation

Digital Communicators:
- Line seizure capability
- Automatic reset
- Abort capability
- Multiple input/output channels for identifying different emergencies
- Listen-in feature
- Monitoring of the alarm system (supervised circuits, telephone line connection, low battery, etc.)

Do-It-Yourself Installation

Purchasing Access Panels - You have a few choices when purchasing these panels. You have to decide upon inside or outside panels and whether you want to operate them with a key or with a digital key pad. Other considerations include alarm system status lights and built-in emergency panic switches.

If you purchase outside panels, make sure they have tamper switches and are designed for outdoor use. In addition, access panels have to interface with the master control panel. Some control panels might not work with a different manufacturer's access panels. Always check the specifications for both the access panel and control panel before spending your hard earned money.

Purchasing Transmitters & Receivers for Wireless Systems - First, find out how many sensors and switches can be connected to a transmitter. Do this by checking the specifications for the unit. Then determine how many transmitters and receivers you will need and whether they should be N.O. or N.C..

You also have to consider the maximum distance a signal must travel to get from a transmitter to the receiver which is usually located in or near the master control panel. Don't purchase a transmitter that cannot relay an alarm message the required distance.

INSTALLING THE ALARM EQUIPMENT

Before trying to install and wire the alarm components, assess your mechanical and electrical skills. Be sure in your mind that you can complete the job before starting it. Always follow the manufacturer's installation and wiring instructions and comply with your local building code requirements. Don't be afraid to consult and hire a licensed electrician or professional installer if you encounter difficulties.

Doing the job right pays dividends of lower maintenance costs and fewer false alarms while increasing the effectiveness of the alarm system.

TOOLS FOR ALARM SYSTEM INSTALLATION

Drill Bits - for drilling holes and starting screws.

Electric Drill Gun - for drilling holes and driving screws

Extension Bit w/ Flexible Shaft & Wire Retriever for drilling holes up through door and window headers, to install magnetic and plunger switches. Also used for pulling wires through walls.

Flexible Shaft Guide - for directing the flexible shaft of the extension bit through walls.

Olm and Volt Meters or Multimeter - for checking continuity, tracing shorts, determining resistance and measuring voltage.

Screw Drivers - for driving screws, mounting and connecting the master control panel, access panels, switches sensors and alarm devices.

Soldering Iron - for making solder joints when splicing and connecting wires.

Wire Cutters - for cutting and stripping wires.

Wire Snake - for pulling wires through walls.

Wire Stapler - for fastening wiring to walls and joints.

SUPPLIES FOR ALARM SYSTEM INSTALLATION

Caulking & Silicone Sealer - to fill and seal holes around wires.

Do-It-Yourself Installation

Electrical Terminals & Connectors - to connect wires to the master control panel and for other wire connections.

Electrical Tape - to wrap & protect wire splices.

Solder - to secure wire splices and connections.

Super Glue or Epoxy Glue - to attach magnetic switches to metal window frames.

Toggle Bolts & Screw Anchors - to secure sensors when mounting them on walls and ceilings.

Wire - to connect the switches, sensors, panels and alarm devices.

Wire Staples and Insulated Wire Nails - to fasten wires to walls, floor and ceiling joists.

Wood & Sheet Metal Screws - to attach switches, sensors, the control panel and alarm devices.

INSTALLING THE CONTROL PANEL, The Automatic Telephone Dialer and the Digital Communicator

You usually install the control panel, dialer and the communicator in the same general location. Install them out of sight in a protected area. You want to be able to get to them without too much trouble, but you do not want them visible. Under a stairway, in a secluded closet or in the basement are good locations. The idea is to keep a burglar from finding them before the local alarms have sounded and the alarm messages have been sent.

In addition, it helps to find a location near the center of the house. This shortens wiring runs for hard-wire systems and signal transmissions for wireless systems.

Mount the master control panel by first finding the wall studs. Then hold the panel up to the wall and mark where the

Home Security Systems

mounting screws should go. Once the panel is mounted and the switches, sensors and alarm devices have been installed, connect them with hard-wiring and wireless transmitters.

The master control panel will need a 120 volt power source. Some panels have a built-in transformer while others require an external transformer to step down the voltage. Either way, find out what the power requirements are for the master control panel before making the connections. Any improper connection could damage or destroy the alarm components as well as cause you injury. Remember, if you have any doubts about the connections, consult a licensed electrician or professional installer.

Connect the master control panel or transformer directly into the electrical circuit rather than by plugging it into a wall outlet. This prevents the panel from being unplugged accidentally. (You may need a licensed electrician to make this connection.) If you have to use a wall outlet make sure it is not controlled by a switch. In addition, make sure the panel is properly ground.

CAUTION - *Always make sure the electrical power is off before making any of the connections.*

The combination master control panel/area sensor should be installed according to the area sensor. That is, install the unit where you need the area protection. Follow the procedures for locating the area sensor, but make sure the control panel is accessible. You might have to place the combination unit where it is visible for the area sensor to work properly. Some manufacturers have recognized this problem and have disguised their combination units to look like stereo speakers which fit nicely on a book shelf.

An automatic telephone dialer or digital communicator is usually installed next to the master control panel in a concealed location. Telephone dialers and digital communicators need telephone lines and telephone jacks install there as well. Your telephone company may permit connections made to a connector block, but could require that a jack be used for either the telephone dialer or digital communicator. (However, line seizure can be accomplished when a RJ 31X jack is properly installed.)

Do-It-Yourself Installation

Always follow the installation instructions supplied with the telephone dialer or digital communicator. Do not connect them to a party line or they might not be able to dial out. It is very important to make the proper connections to the telephone lines so that neither your components nor the telephone company equipment is damaged. The telephone company may insist on installing the jacks and making the connections for you.

Installing Access Panels - By now, you probably have decided upon inside or outside access panels and where they are going to be installed. The next step is to determine exactly where on the wall to place them. Depending on your particular panels, you will probably fasten them to a wall stud. Find the studs and locate any electrical wires and utility pipes before cutting into the wall. Also, determine how and where you are going to run wires to the master control panel.

Cut into the wall and install the panel box. (Some manufacturers include a template for exact hole size.) Then pull the wires and connect them to the master control panel and the access panel. For outside panels, connect the tamper switch to an non-delay protection circuit.

Installing the Local Alarms - It is a good idea to install at least one siren or bell inside and at least one siren or bell outside your home. The outside alarm scares the intruder away as he tries to break-in. It does this by attracting attention while alerting neighbors to call police.

The inside alarm is to scare off the intruder if he gets inside before the alarm sounds. (This can happen if he comes through a door on a delayed circuit.)

Install the inside local alarm away from the master control panel so if the burglar or intruder searches for it hoping to disable the system, the sound will not draw him to the panel. Hide the bell or siren so an intruder cannot tell where the sound is coming from. Behind false ceilings and air vent grills are good installation locations.

Install the outside alarm where it cannot be reached and disabled by a burglar. You might put the siren or bell behind an attic vent which faces the street. This placement hides and protects the alarm but could reduce its sound output. Always make sure it can be heard by a neighbor who will call police.

Home Security Systems

Another idea is to mount the outside alarm in an alarm box. The box protects the alarm from the elements and from burglars. Install a tamper switch in the box so if a burglar tries to disable the siren or bell, an alarm will be initiated.

Some homeowners want the alarm bell or siren to be seen. They believe its presence will deter a burglar before he tries to break-in. Others place the alarm bell or siren out of sight thinking that by surprising the intruder the noise will cause him to panic and make a fast exit.

INSTALLING THE PERIMETER SWITCHES & SENSORS

Magnetic Switches - Doors - Install the recessed or concealed magnetic switches by drilling a hole for the switch portion into the top of the door frame. This hole is drilled up through the header and into the attic. Locate the wall studs and wiring before drilling. (This is extremely important because otherwise you could damage other wiring and injure yourself.)

By using an extension drill bit with a wire retriever, wires can be pulled back through the hole from the attic. Connect the switch by soldering the wires and then use electrical tape to wrap and protect the connection.

Next, push the switch portion back into the hole. Make sure that the switch fits flush with the door frame. Secure it with an adhesive or with small nails. Drill a hole for the magnet into the top edge of the door, checking to make sure it aligns with the switch. Then measure the gap between the switch and magnet, making sure it does not exceed the manufacturer's specifications. Follow the installation instructions provided and remember to install magnetic switches opposite the door hinges so that even the slightest opening activates the alarm system.

Surface mounted magnetic switches are also installed opposite the door hinges. They are usually attached to the door and door frame with screws. As with flush mounted switches, be careful to follow the gap or spacing requirements between the switch and magnet.

In a hard-wire system you can conceal the wires behind door molding and trim pieces. As an alternative to hiding the

INSTALLING RECESSED MAGNETIC SWITCHES

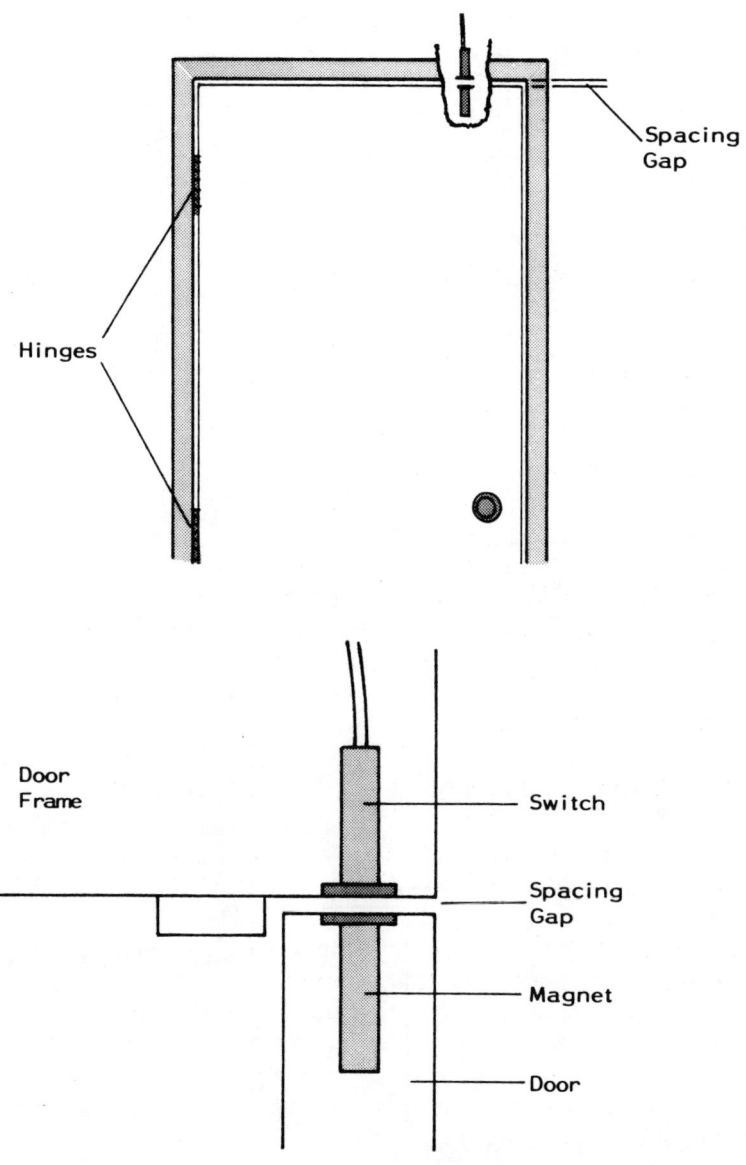

INSTALLING SURFACE MAGNETIC SWITCHES

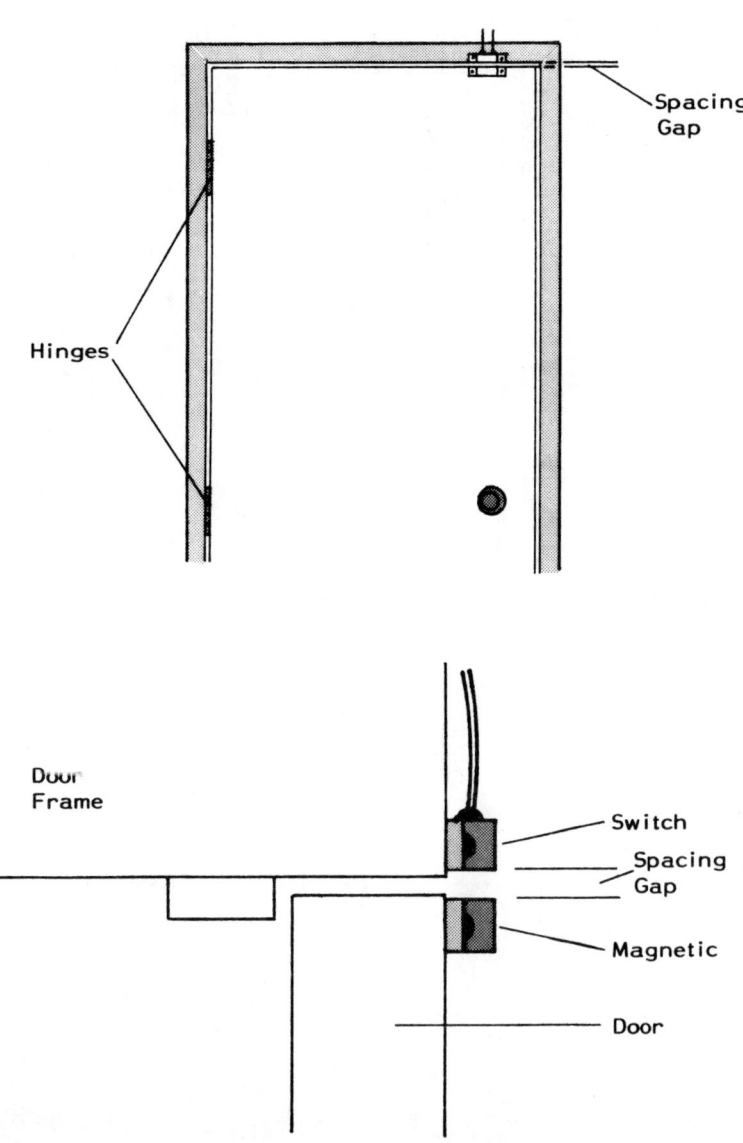

Do-It-Yourself Installation

SURFACE MOUNTED MAGNETIC SWITCHES
(as viewed from inside)

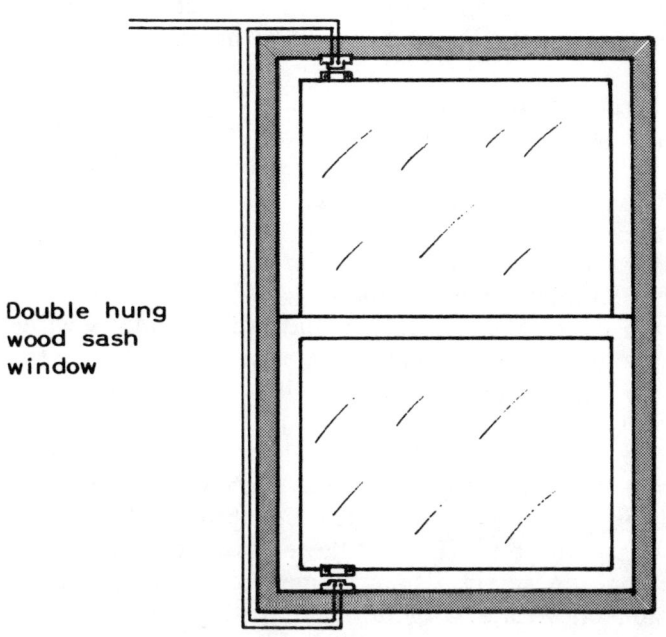

Double hung wood sash window

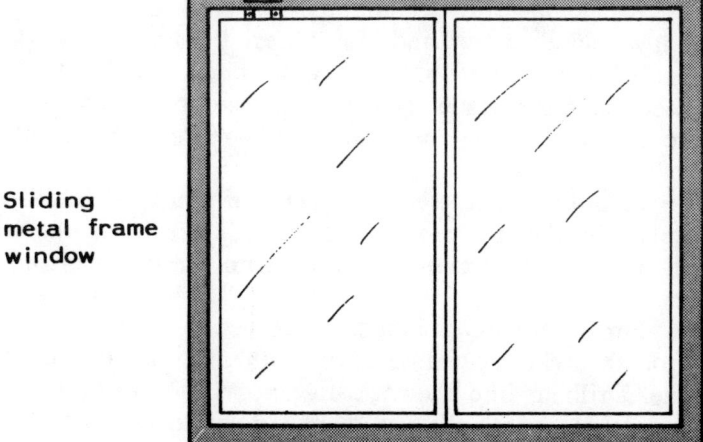

Sliding metal frame window

Home Security Systems

wires behind molding and trim, try using flat and colored wires that blend with your wall coverings.

If you use wireless transmitters, surface mounted switches will be easier to install and connect. Some wiring will be exposed, but the installation will be much faster and easier.

Magnetic Switches - Windows - Flush mounted magnetic switches can be used on most types of windows. However, some metal frame windows may require a specially designed switch to work in metal frames and some metal frames do not provide enough space to mount a concealed switch. It is usually easier for a do-it-yourselfer to install surface mounted switches on metal frame windows.

On double hung windows and other windows with wood frames and sashes, use the same recessed magnetic switches used for doors. Install the switches by the same methods previously described.

Double hung or double sash windows will need two separate switches: one for the upper sash and one for the lower sash. For the upper sash, drill up through the frame and header into the attic. For the lower sash, drill down through the frame and sill into the basement or crawl space. Use an extension bit with a wire retriever to drill the holes and pull the wires.

Be careful when drilling holes for the flush mounted magnets especially in double pane windows. It is very easy to drill into the glass and break it.

Attach the surface mounted switches to the wood frame windows with screws and use super glue or epoxy glue for metal frame windows. As with doors, hide the wires connecting the surface mounted switches behind molding and trim. This makes your installation look more professional.

Installing Recessed Plunger Switches - Install these switches in the door jambs on the same side as the hinges. Drill into the jamb and then down into the basement or crawl space.

For double hung windows, locate a plunger switch in the top of the window frame and another in the bottom of the frame. Drill up into the attic to run wires for the top switch and drill down into the basement for the lower switch. Again, follow the drilling precautions described above.

Do-It-Yourself Installation

Installing Glass Breakage Sensors - Installing these sensors is easy. Most are self-sticking so be sure to clean the glass before applying them to the inside corners of the window panes. Follow the manufacturer's instructions for exact placement and wiring.

Install a breakage sensor in every window pane that is large enough for an intruder to come through after he breaks out the glass. Large windows may need more than one sensor so find out how much surface area a breakage sensor will protect before buying.

Some breakage sensors can be adjusted to work on either a N.O. or N.C. circuit, but most will be connected to a N.C. circuit for perimeter protection. With some breakage sensors you might need a pulse stretcher or fast response circuit in the master control panel. A pulse stretcher and fast response circuit react to the very short duration of the alarm signal generated by the breakage sensors. Without one of these, an alarm signal may be missed.

Finally, adjust the sensitivity so that low flying aircraft, sonic booms and other vibrations will not trigger a false alarm.

Installing Shock Sensors - Mount the sensing unit on the window frame according to the manufacturer's installation instructions. Shock sensors protecting other windows can be wired in series and then connected to the same protection circuit. This saves time and money on the installation especially when using wireless transmitters.

Installing Window Foil - The first step is to clean the glass so the foil will adhere to it. Then use chalk or a grease pencil to mark guide lines on the outside surface of the glass three inches out from the frame. Mark a line all the way around the window.

Apply a continuous strip of self-sticking foil to the inside glass surface following the guide lines. Stretch the foil very slightly and smooth it as you go being careful not to break or tear it.

When you come to a corner, fold the foil in the opposite direction you want to go and then fold the foil back onto itself. This produces a clean 90 degree corner without cutting or splicing.

INSTALLING GLASS BREAKAGE SENSORS

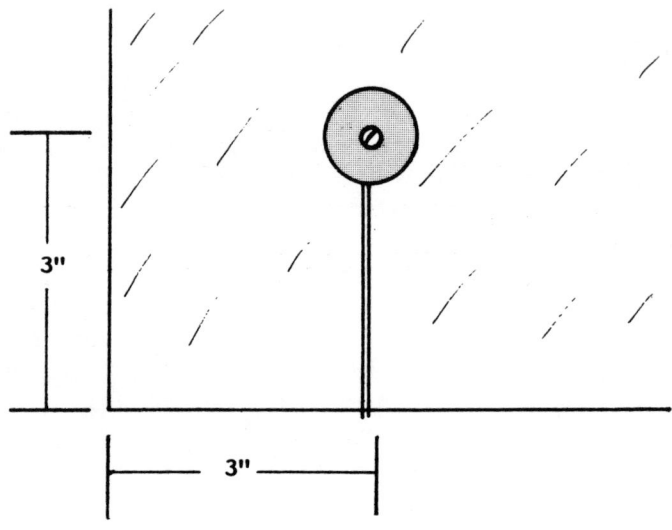

INSTALLING SHOCK SENSORS
(as viewed from inside)

INSTALLING WINDOW FOIL
(as viewed from inside)

TAKE-OFF CONNECTOR

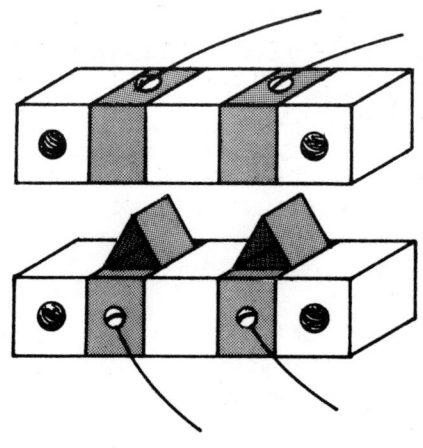

Sliding Door or Window Installation

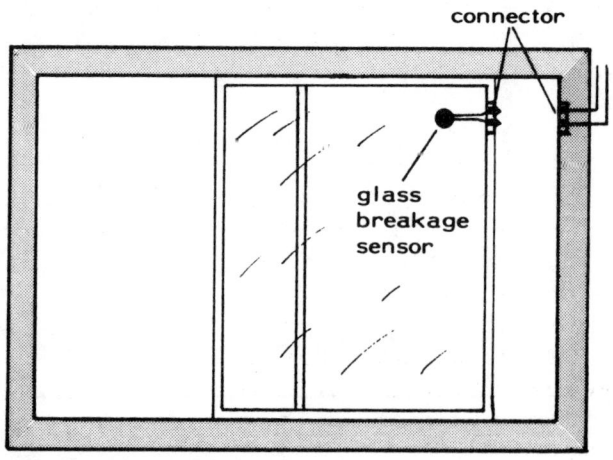

Do-It-Yourself Installation

Finally, install the foil connector blocks on the windows and attach the foil. Make sure that neither the foil nor the connectors touch a metal window frame as it could short out the protection circuit

In addition, you should varnish over the foil to protect it. Without this protection, the foil could be damaged quite easily and create false alarms. But do this after you have tested the circuit and determined that the foil is not cracked or making a bad connection.

Window Take-Off Connectors - Sliding windows and sliding glass doors present problems when installing glass breakage sensors and window foil. The window or door cannot be opened without disconnecting the sensor or foil from the alarm circuit wires.

One solution that allows protected windows to be opened is to install take-off connectors. First, a sensor or foil is applied to the glass. Next, one portion of the two part connector is attached to the movable frame and wired to the sensor or foil. The other part of the connector is then mounted to the window sill or stationary frame. When both parts of the connector are aligned, the breakage sensor or foil can be "unplugged" to open the window.

With take-off connectors, always remember to close and lock the door or window before trying to arm the alarm system. Leaving an open circuit could prevent the system from arming or may initiate a false alarm.

Wired Window Screens - Mount the screens to the windows using the mounting hardware provided. Then connect the wires to a normally closed protection circuit using a connector plug. This plug allows you to completely remove the screen for window cleaning.

Some wired window screens come with flush mounted magnetic tamper switches implanted into the bottom of the frame. Actually, the magnet is hidden in the frame while the switch portion is recessed into the sill concealing the switch from burglars.

If your screens do not come with tamper switches, install a normally closed surface mounted magnetic switch at the bottom of the frame and attach the magnet portion to the sill.

WIRED WINDOW SCREEN INSTALLATION

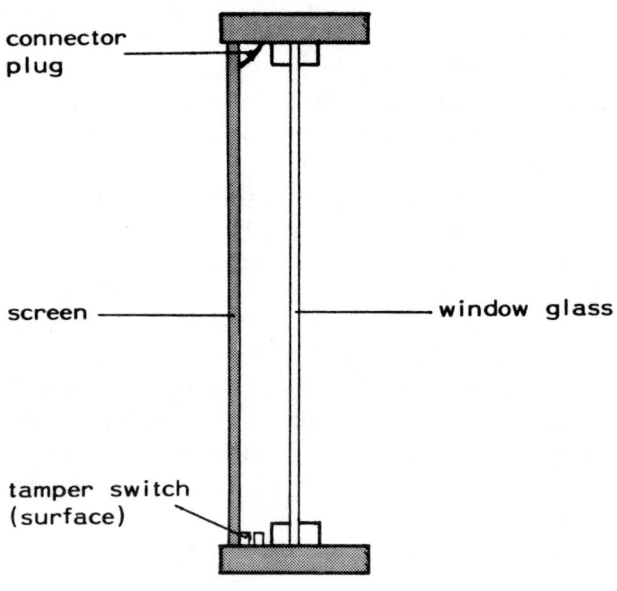

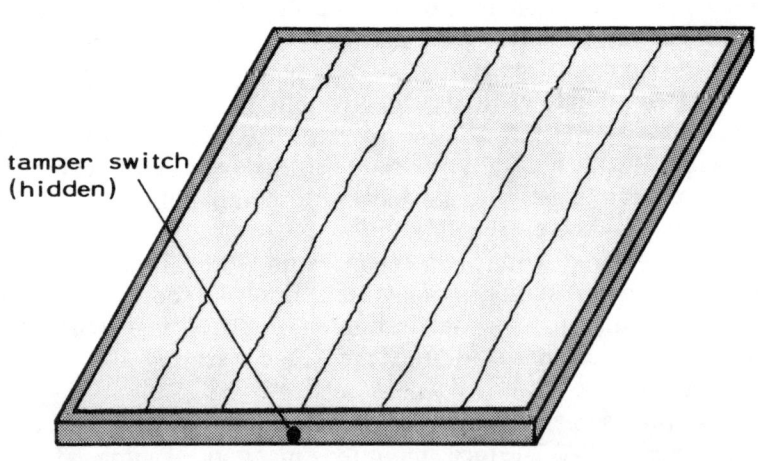

Do-It-Yourself Installation

When this homemade tamper switch is connected to a normally closed circuit, the installation is complete.

Whether your wired window screens come with flush mounted magnetic tamper switches or if you have to install your own surface mounted switches, make sure an alarm will be initiated if a burglar tries to remove one of the screens.

Sound Discriminators & Audio Detectors - Install the discriminators or audio detectors near potential entry points. One sensor can protect both doors and windows in a room. But make sure that the total room area does not exceed the range and capacity of the sensor. After connecting the sensors to the protection circuit, connect the power source remembering to follow the manufacturer's installation and wiring instructions.

If you decide to install a sound discriminator that uses a line carrier system, the installation is simple. First, locate the master control panel then plug the sensors into wall outlets in the areas and rooms you want to protect. Finally, test and adjust the sensors.

INSTALLING AREA PROTECTION SENSORS & DETECTORS

Follow this general sequence when installing area protection sensors and detectors:

1) Determine the areas or rooms you want to protect by referring to your floor plans. Find the places where a burglar would have to enter or pass through as he searches for valuables. It is not necessary to install an area sensor in every room, especially if you intend to install a combination perimeter/area protection system.

2) Mount the area sensors according to the manufacturer's recommendations and instructions for exact placement on walls and ceiling.

3) Connect the sensors and detectors to the master control panel by hard-wiring or by wireless transmitters.

Home Security Systems

4) Connect the sensors to their power sources paying strict attention to the voltage requirements and wiring instructions.

5) Test and adjust the sensors and detectors so they will cover the areas and rooms but not spill into other areas where they could trigger false alarms.

Most microwave, ultrasonic and infrared sensors have a "walk test" light which is used to test and adjust the sensor. First, switch the alarm system into the test mode and then walk through the protected area watching for the test light to come on. After a few passes you will know just where the sensor will detect an intruder. Start by setting the sensitivity on the lowest level and then adjust the sensor outward until you are satisfied that it will cover the area. On those sensors that cannot be adjusted, use the test light to make sure the sensor has enough range to cover the protected space.

Microwave Installation - Mount the sensor and then connect it to the master control panel with hard-wiring or with wireless transmitters. Next, connect the power supply and use the "walk test" light to adjust the sensor. Turn the adjustment setting to the lowest level and adjust the pattern outward. Keep testing and adjusting the pattern until you are sure that the microwaves are contained and do not spill outside the house.

Ultrasonic Installation - Mount the sensor and connect it to the master control panel. Connect the power source and adjust the sensor. This sensor has both a range and sensitivity adjustment. Start by adjusting the range using the "walk test" and then adjust the sensitivity level.

Passive Infrared Installation - Mount the sensor and connect it to the protective circuit in the master control panel. Then connect the power supply. These sensors have no range adjustment, but their angle of coverage can be adjusted to cover a room or space. Remember, passive infrared detectors are more sensitive when an intruder walks across the protected area, so mount and adjust the angle keeping this in mind.

Photoelectric Beam Installation - Mount the transmitter-receiver and reflector high enough so that an intruder won't step over it but low enough so that he won't pass under it. The idea is to make sure he will break the beam and initiate an alarm.

After mounting the transmitter/receiver, connect it to the protective circuit in the master control panel and then connect the power supply. Mount the reflector and align it with the detector. Finally, test the detector to make sure it works properly.

Installing Mat Switches - First, take up the carpet in those rooms or hallways where the mats are going to be installed. It is not necessary to completely remove the carpet. Sometimes you can remove one edge and slide the mat under it. Make sure the mats you install are large enough so an intruder will not step over them.

Connect the wire leads from the mat switches to normally open circuits and replace the carpeting.

Sometimes you need to connect a mat switch to a normally closed circuit to interface with a particular protection loop or zone. In this case, converters are available which change a normally open alarm signal into a normally closed signal.

> **IMPORTANT** - *Always follow the manufacturer's instructions and procedures for placement, wiring connections and power requirements when installing any of these sensors.*

Home Security Systems

SOME DON'TS WHEN INSTALLING AREA SENSORS

The following is a list of things you should watch out for when installing area sensors:

Microwave
1) Don't point the sensor at a window or outside walls that will be penetrated by microwaves. An alarm could be initiated by someone or something on the outside of your home.

2) Don't point the sensor towards a ceiling fan that starts and stops automatically.

3) Don't use microwave sensors near fluorescent lights.

4) Don't allow large pets into a room or area protected by a microwave sensor. They could initiate an alarm.

Ultrasonics
1) Don't install the sensor in an area where any pets could pass through. Even small pets can create false alarms with ultrasonics.

2) Don't point the sensor at drapes that will move when a heater or air conditioner comes on.

3) Don't put a telephone that will ring in a protected area. The ringing changes the sound wave patterns and can trigger the alarm system. One solution is to turn the ringer off or take the bells out of the phone.

Passive Infrared
1) Don't place a sensor in an area subject to rapid temperature changes.

Do-It-Yourself Installation

2) Don't locate a sensor where heater or air conditioner drafts will cause enough of a temperature change to initiate an alarm.

3) Don't install a sensor near a refrigerator. The motor starting and stopping can create just enough of a temperature change to set off the alarm system.

4) Don't allow sunlight or headlights to enter the protected area. They may cause a slight but rapid temperature change creating false alarms.

Photoelectric Beam
1) Don't allow sunlight or headlights to hit the receiver or reflector.

2) Don't use beam sensors in areas where furniture or other movable objects will interfere with the light beam getting from the transmitter to the receiver.

3) Don't allow pets into the protected areas when the alarm system is armed.

4) Don't exceed the manufacturer's maximum beam distance when trying to protect an entire room.

INSTALLING PANIC SWITCHES
& Smoke and Heat Detectors

Panic Switches - Install a panic switch near your front door so if an intruder tries to force his way in, you can sound the alarm immediately. In addition, you might want to install other panic switches in your bedroom and near the center of your house. You want to be able to get to a switch quickly in case an intruder gets inside while you are home.

Mount one switch in the wall next to your bed and another in the wall next to your entry door. Locate the

Home Security Systems

existing wiring and wall studs before cutting out a hole for the switches. Use a flexible extension bit and wire retriever to pull wires up from the basement or down from the attic. Then make the connections to the 24 hour panic circuit in the master control panel.

Smoke Detectors - The total number of smoke detectors you install depends on the size and layout of your home. One detector is better than none and two are better than one. The idea is to be warned as soon as a fire starts no matter where it is in your home. Also, if one detector fails another will sound the alarm.

Consider installing both photoelectric and ionization detectors in your home to get the advantages of each.

As a minimum, install a smoke detector between the sleeping areas and the rest of the house and at least one detector on every level of a multi-level home. It is very important that the fire alarm can be heard throughout the house whether you install self-contained detectors or detectors that connect to the master control panel.

For even more protection, install detectors in each bedroom, in main hallways, living rooms, family rooms, basements and at the top of every stairway. The point is to install as many detectors as practical. Locate them where smoke will be detected as soon as a fire breaks out and before it reaches the bedrooms.

The precise locations of the detectors is very important. Mount them high on the wall or on the ceilings - remember smoke rises. But watch out for thermal barriers which restrict smoke movement. These barriers occur in corners where walls and ceilings meet. Air does not circulate well in those places. Do not install detectors near heater or air conditioner vents or near air return vents. Smoke could be blown away from the detectors.

Placement is also important in preventing false alarms. Do not install smoke detectors in dusty or dirty areas or in damp and excessively humid locations. Dust and water droplets can fool the detectors and initiate an alarm. In addition, be careful about placement in or near kitchens as cooking smoke will cause the detectors to react.

Do-It-Yourself Installation

If you use smoke detectors that connect to a protection circuit in the master control panel, make sure that circuit is supervised and activated 24 hours a day.

> **IMPORTANT** - *Always follow the manufacturer's instructions for placement and connection of your smoke detectors. Also, you must maintain the detectors. It is essential to check and replace the batteries in self-contained detectors and the standby batteries in the master control panel.*

Heat Detectors - Install heat detectors in areas where smoke detectors will be subject to false alarms. These areas include kitchens, attics, garages and furnace rooms. Connect them to the same supervised fire protection circuits as the smoke detectors.

To repeat, heat detectors should be used *with* and not instead of smoke detectors.

Both smoke and heat detectors are only part of a complete home fire safety program. That safety program should also include fire prevention, fire extinguishers, evacuation plans, escape equipment and periodic fire drills.

HOW TO WIRE YOUR ALARM SYSTEM

When connecting switches and sensors to the master control panel, there are three basic methods to choose from: hard-wiring, wireless transmitters and line carriers. You can use one method or a combination of all three, but almost all alarm systems will need some hard-wiring. Even with wireless alarm systems, switches and sensors have to be hard-wired to the transmitter.

Hard-wiring is also used to connect alarm devices and remote access panels to the master control panel in wireless systems as well as in line carrier alarm systems.

In this section you will see the basics of connecting the alarm components by installing wireless transmitters and by using hard-wiring.

Home Security Systems

Wireless Transmitter Systems - Wireless transmitters and receivers save a great deal of installation time and effort. Sometimes wireless transmitters are a necessity because of restricted access in the attic or under the house. This is especially true if your house is built on a slab foundation and does not have an attic.

The first thing to do when installing wireless transmitters is to locate and mount all of the switches and sensors. Then mount the transmitters and make the connections being sure you connect only normally closed switches and sensors to transmitters designed to accept a N.C. switch or sensor. Also, only connect sensors designed for normally open circuits to transmitters which will accept a N.O. signal.

You can save money by connecting several switches or sensors in a room to one transmitter. But don't connect more switches and sensors than the transmitter is designed to handle.

If you install a zoned system, be careful to connect switches and sensors to the proper transmitter so the receiver and control panel will know what the alarm is and where it is coming from.

Hard-wire Systems and Hard-wiring - Hard-wiring switches, sensors and alarm devices to the master control panel is usually done with low voltage wire, allowing the homeowner to make the connections without hiring a licensed electrician. But always consult your local building code for low voltage wiring requirements. There is debate in some communities as to whether low voltage alarm wiring is subject to the same high voltage regulations.

When you hard-wire an alarm system, special tools make the job much easier. Long flexible bits with wire retrievers are available. Use them to drill up through door and window headers into the attic and to drill down through sills and sole plates into the basement. After drilling through a wall and into the attic or basement, attach the wires to the drill bit wire retriever and pull them back through the hole. Make the connection to the switch or sensor and then fill the hole around the wires with caulking.

As an alternative to drilling through walls, some installers run wires under carpeting in those areas not subject to foot traffic. They will run wires next to walls to prevent them from being stepped on. However, this method of running wires

WIRELESS TRANSMITTER SYSTEM

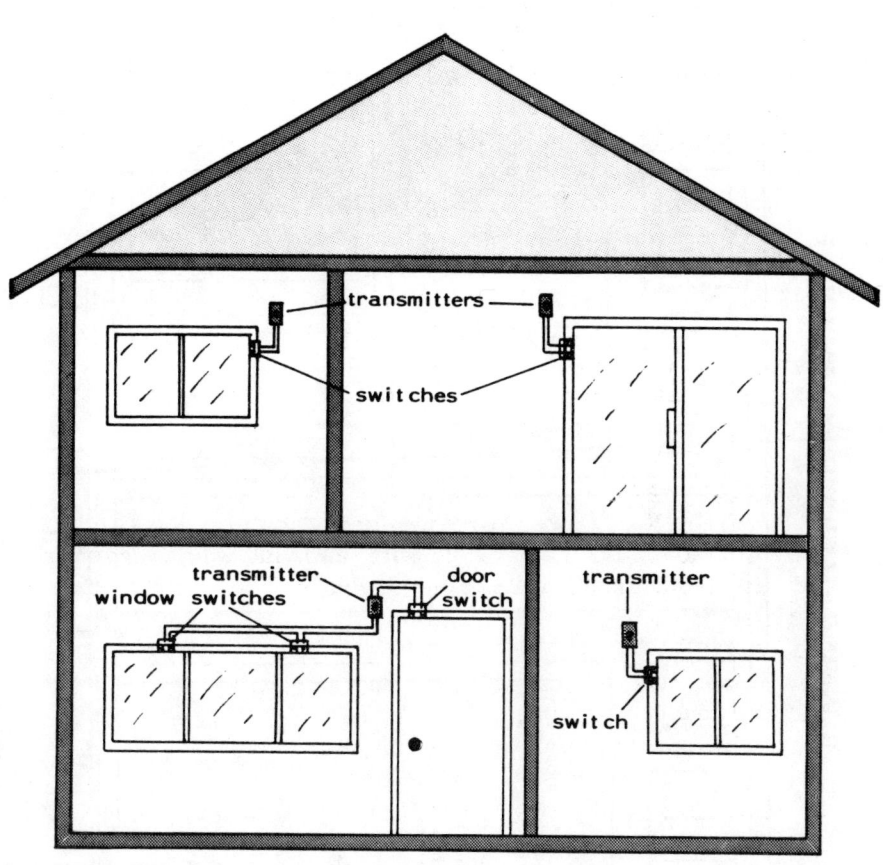

RUNNING HARD WIRING THROUGH WALLS

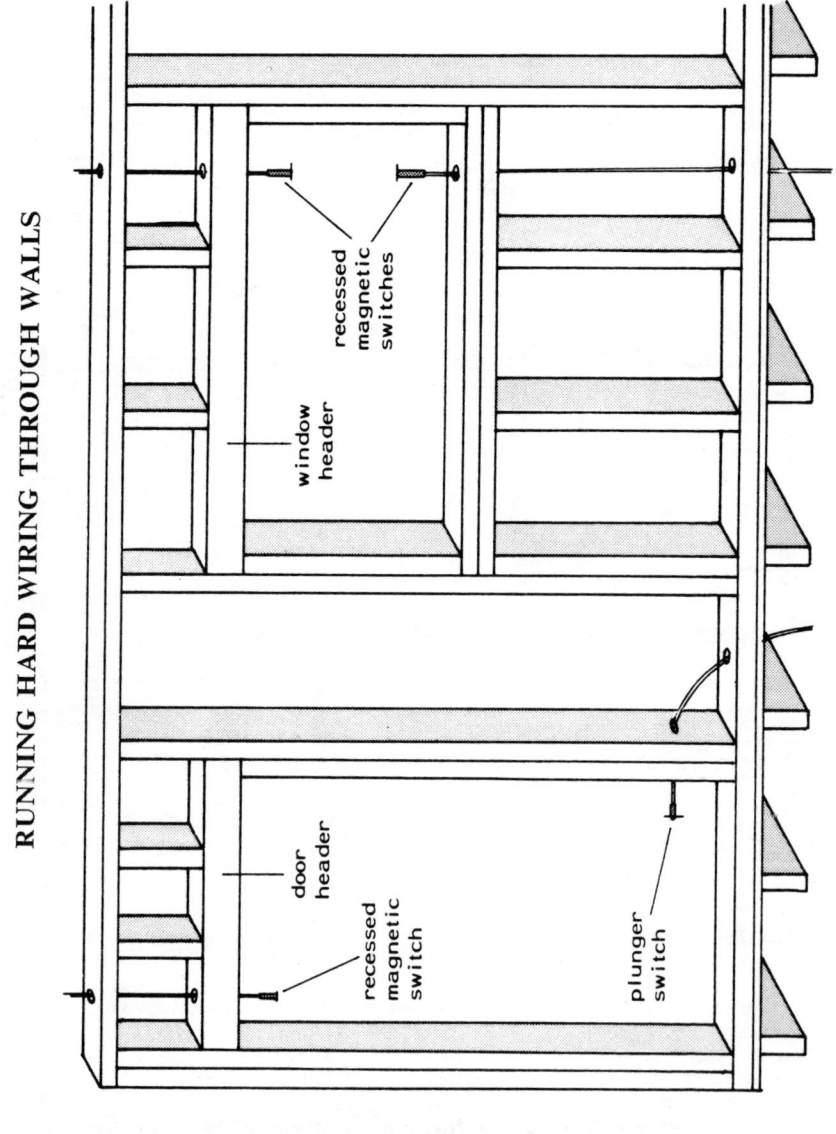

REPRESENTATIVE WIRING DIAGRAM
NuTone S-2252 Security System

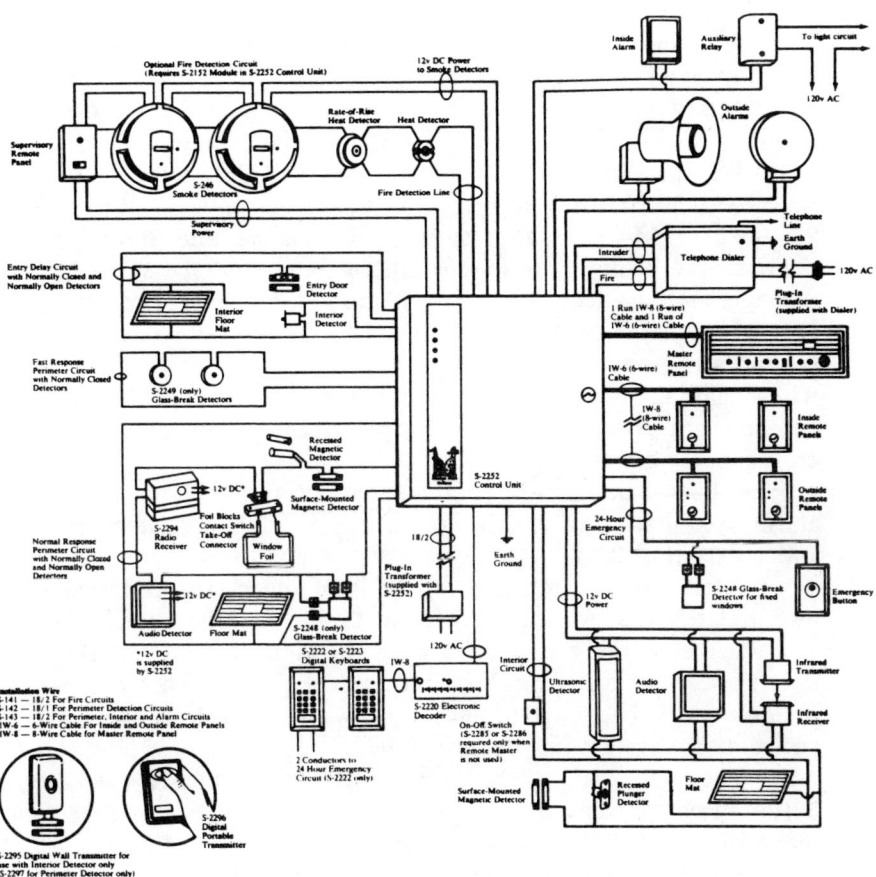

This diagram illustrates the type of wiring needed for a comprehensive NuTone Security System. It is provided simply to assist you in your initial system planning. It is not intended for actual installation, nor does it include all devices that might be used. Always refer to specific installation instruction sheets that are enclosed with each system for complete wiring information.

Diagram courtesy of NuTone, a Division of Scoville.

Home Security Systems

under carpets in not recommended because it can be unsafe and may violate local wiring codes.

If your wiring has to be run in outside or exposed areas, consider using rigid conduit to protect it from the elements as well as intruders.

You should also refer to a wiring guide for voltage drops over a given wiring run. The wire size might have to be increased where the runs are excessively long. If in doubt about wires or wiring, consult a licensed electrician.

When wiring components, always follow the manufacturer's wiring size recommendations. (Usually 18 AWG to 22 AWG) Make as few splices as possible and solder the connections. Make your splices where they won't be subject to damage and use two layers of electrical tape for protection. Splices in damp areas should be sealed to prevent moisture from damaging the connection. Follow the correct polarity when making connections to the control panel and the other components. Also make sure the panel and other components are properly ground.

These precautions reduce potential problems in the alarm system wiring which can save you a great deal of time and frustration later.

After each switch and sensor is mounted and connected, test it and the wiring before installing another switch or sensor. If something is not working the time to find out about it is now. Nothing is more irritating and time consuming than to look for a bad switch, sensor or wire connection after the entire alarm system has been installed.

The less visible the wiring and components are the better. Try to conceal as much of the wiring as possible to make the job look more professional.

Finally, secure the wires to walls, floors, ceilings and support joists with insulated brads and wire nails. Install a brad every two feet to keep wires in place.

IMPORTANT - *Locate the existing wires and utility pipes before drilling into walls, floor and ceilings.*

Do-It-Yourself Installation

TESTING YOUR ALARM SYSTEM

When you have finished installing your alarm system, test it. Check all the door and window switches by opening the doors and windows to initiate an alarm. Test the area sensors by walking through the protected spaces. Check all the functions of the master control panel, access panels, telephone dialer and digital communicator.

Also, make sure you and your family have a thorough knowledge of how your alarm system works. You should know how to operate it and be able to recognize when a component is not working properly.

Another important thing to do is to tell your neighbors that you have an alarm system. Set off the local alarms so they will know the sound. Instruct them to call police or the fire department if they hear the alarms and provide them with the phone numbers of both the police and fire departments as well as your office number. Also, give them the phone number of a friend or relative who can contact you in an emergency.

Finally, it is very important not to talk about your alarm system or your valuables even to friends and neighbors. You cannot control who they talk to or what they will say. The less other people know about your alarm system the better. This alone is a very good reason to install the alarm system yourself.

PERIODIC TESTING & MAINTENANCE

Test your alarm system periodically to insure it will work properly if needed. Check all switches and sensors as well as the master control panel and access panels. Test the local alarms, automatic telephone dialer and digital communicator. But remember to tell the neighbors and the central station monitor *before* you set off the alarm so they will know it is only a test.

Batteries should also be tested and replaced periodically. In addition, clean the sensors, inspect the wiring and check sensor mountings regularly.

FALSE ALARMS

Hopefully, you have designed and installed your alarm system properly keeping false alarms to a minimum. But some false alarms are inevitable. Make sure you understand your alarm system thoroughly and remember to disarm it upon returning.

False alarms are irritating to neighbors and police. Police waste time and money responding to them while claiming as many as 90% of all alarms are false. Some cities are fighting back by forcing homeowners to reduce false alarms by imposing fines and revoking alarm system permits.

You can help reduce false alarms by:

1) **Designing and installing your alarm system properly.**

2) **Using the appropriate sensors, switches and components.**

3) **Operating your alarm system correctly by remembering to disarm it upon returning.**

4) **Testing your alarm system periodically.**

ALARM SYSTEM WARNING SIGNS

After you have installed your alarm system you might want to advertise that fact to potential burglars and intruders. Alarm window stickers and yard signs can deter them from trying to gain entry. Why should a burglar or intruder risk detection and possible apprehension? A house without an alarm system would be much easier to enter and burglarize.

Your alarm stickers and yard signs should only indicate that your home is protected by an alarm system and not identify any of the equipment by name or manufacturer.

Place the window stickers on rear windows as well as the front windows. You want to make sure a potential burglar or intruder will see one.

ALARM SYSTEM WARNING SIGNS

WARNING!
THIS HOME IS PROTECTED BY AN ELECTRONIC ALARM SYSTEM

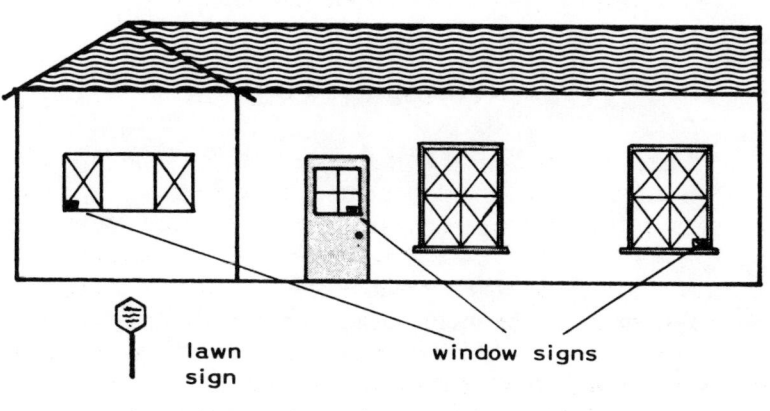

Chapter 10

Professional Alarm Systems

The previous chapters have shown you how to determine your security needs, how to plan an alarm system and finally, how to install the components.

If installing your own alarm system seems too complex or if you simply do not have the time, this chapter is for you. You will learn how to shop and compare security services as well as equipment offered by professional alarm dealers and installers.

Most professional alarm companies can design your system, provide all the equipment, arrange central station monitoring and service the system.

If you had trouble determining your security needs or if you were unable to decide how to protect your home, you could hire a security consultant. He will determine your security needs by the same methods described in Chapter 8. In addition, he will use his knowledge of alarm equipment and his experience to design a system to fit your needs.

An alarm company usually provides this planning service before they install a security system. Many companies will make a preliminary survey of your home for free, provided they get a chance to sell you their services.

Get written estimates from these alarm companies detailing all the equipment to be installed, all the work to be performed and all the services to be provided. Be sure the estimates specify who will obtain and pay for any permits or licenses that may be required: the alarm installer or the homeowner.

Home Security Systems

Getting the most for your money is more than accepting the lowest bid. Value in a home security system is a combination of good equipment, proper installation and prompt response in an emergency.

HOW TO FIND AN ALARM COMPANY

Start with the telephone book by looking under *"Alarms"*, *"Burglar Alarms"* and *"Security Systems"*.

Find out if your area has an association of alarm dealers and installers. They might be able to recommend an alarm company which adheres to the association's standards for alarm system installation and monitoring. Contact the National Burglar and Fire Alarm Association (NBFAA), if you have trouble finding an association in your area.

In addition, you can ask your police and fire departments for referrals and recommendations. Also ask your insurance agent if he or she can recommend an alarm company.

One of the best sources is a referral from a satisfied customer. Ask your friends and neighbors who have installed alarm systems what their experience has been with different alarm companies. Ask if they have had any problems with the equipment, the installation, monitoring service or with false alarms. Also find out if they have had any alarm conditions where either the police, the fire department, paramedics or private guards were dispatched and how long the response took.

Be suspicious of door to door alarm system salesmen. It is not very smart to let a stranger into your home to analyze your security needs. It is possible that he is checking your house out only to burglarize it later.

This is why an established alarm company should be contacted. But even large and established alarm companies can be infiltrated by a dishonest employee.

My purpose is not to scare you, but to make you aware that you have to be extremely careful when dealing with alarm companies and installers. It is unfortunate that one bad company or dishonest employee does a major disservice to the entire alarm industry.

Professionally Installed Alarm Systems

HOW TO CHOOSE AN ALARM COMPANY

Contact the Better Business Bureau to find out if the alarm company you are considering has had any complaints filed against it. Also check with consumer protection agencies in your city or state for similar complaints.

Be very cautious when dealing with a new alarm company. The company will not have had time to develop a track record either good or bad. In addition, make sure they will be able to complete the installation once it is started. Do this by requiring a completion or performance bond.

Dealing only with established alarm companies with numerous satisfied customers makes good sense. Check out any company thoroughly before signing a purchase and installation contract.

It is important to select the right alarm company to install your alarm system. Here is a list of questions to ask:

1) How long has the company been in business?

2) How many alarm systems have they installed?

3) What types of installations do they specialize in? Commercial, industrial or residential?

4) What services can they provide: security consulting, system design, installation, central station monitoring, service and repair?

5) Are their employees bonded and have they been screened for criminal records?

6) What trade associations do they belong to?

7) Are they members of the National Burglar and Fire Alarm Association (NBFAA) or one of its affiliates?

Home Security Systems

8) Does the company have the proper city, county and state licenses and permits?

9) Can they give you names and telephone numbers of their customers? (They probably will refuse because of security and liability reasons.)

10) Do they use their own employees or do they subcontract the installation?

11) Do they carry insurance such as errors and omissions, general liability, workers compensation and operations coverage?

12) Will they provide a completion or performance bond?

13) How long will you have to wait before they can install your system?

14) How long will it take them to install the system?

15) How long does the alarm company guaranty the entire alarm system after the installation is complete?

16) Will the company finance or lease the alarm system or will you have to make your own financing arrangements?

17) Do they have their own central station monitoring facility or do they contract it out?

18) What liability does the alarm company assume if the alarm system malfunctions and the homeowner suffers a loss?

Professionally Installed Alarm Systems

Make sure you have a clear understanding with the alarm company before signing a contract. Review their liability limitations and liquidated damages clause. (Alarm companies try to limit their liability by using such clauses.) Also, liability limitations will differ from state to state. Try to remember you are buying an alarm system and alarm services from these companies. You are **not** buying insurance against loss.

UNDERWRITERS LABORATORIES & THE SECURITY INDUSTRY

Underwriters Laboratories of Northbrook, Illinois has made a significant impact on security alarm equipment and its installation. Underwriters Laboratories, or "UL" for short, is the recognized standard for both the manufacturing of alarm equipment and the monitoring of alarm systems.

In relation to UL and the security industry, here are three things to look for when buying alarm equipment and services for your home:

<u>UL Listed Equipment</u> - This means that the equipment has met certain standards and requirements of Underwriters Laboratories. (Again, equipment without a UL listing does not mean that it is not good. But make sure you find out why something does not have a UL listing before buying it.)

<u>UL Listed Alarm Companies</u> - To be a UL listed, an alarm company must use UL listed equipment, installed according to UL techniques. Also, the company must have had previous alarm system installations inspected and passed by UL. Then the company must undergo random inspections of their installations to insure that the UL standards are being maintained.

<u>UL Certified Alarm System</u> - For an alarm system to be UL certified it must be installed by a

UL listed company according to UL standards using UL listed equipment.

While most of the alarm companies and most of the alarm systems are not UL listed or certified, much of the alarm and security equipment on the market is listed by Underwriters Laboratories.

For more details on UL home security system guidelines, refer to UL 1641 titled, *"Standards for Installation and Classification of Residential Burglar Alarm Systems"*.

QUESTIONS TO ASK ABOUT
Components, Installation & Central Station Monitoring

Once you have selected an alarm company, find out how they plan to protect your home. Ask what kinds of protection systems they recommend and what kind of response they suggest for your security needs. Each recommendation will determine what equipment will be installed and what the alarm system will cost to install and operate.

The following is a list of questions to ask the alarm company about components, installation and central station monitoring:

Components

1) What brands do they install?

2) How long has the manufacturer(s) been in business?

3) What has their experience been with the manufacturer and the reliability of the alarm equipment?

4) How long has the alarm company been installing this brand of equipment?

5) How long are the warranties and what is covered: parts, labor?

Professionally Installed Alarm Systems

6) Who services the equipment and how long will it take if repairs have to be made?

7) Are the components listed by Underwriters Laboratories?

Installation

1) How long will the installation take to complete?

2) How will the switches and sensors be installed? Will they be surface or flush mounted?

3) How is the wiring to be done? Is it going to be run through walls, under the house and in the attic or simply run under the carpet?

4) Are wireless transmitters going to be used?

5) Who will do the installation? Are they employees or will the job be turned over to subcontractors?

6) Have the installers had background checks for criminal records and are they bonded?

7) How experienced are the installers? Do they have an installation certificate from a qualified training school?

8) Will the alarm company provide a performance bond guaranteeing that the installation will be completed and that the supplies and subcontractors will be paid?

Central Station Monitoring

1) How long has the central station been in business?

Home Security Systems

2) How many subscribers do they have?

3) Is the station listed for residential monitoring by Underwriters Laboratories?

4) Will the company who sold and installed your alarm system do the central station monitoring, or is it contracted out?

5) What kinds of response will be provided? Police, fire department, paramedics, private security guard?

6) How long will the response take?

7) What kinds of equipment and power backup system does the central station have, to insure that your home will not be vulnerable if their equipment breaks down or if the power fails?

8) Who is liable for a loss to the homeowner if equipment fails or if no response is sent?

9) What is the monthly cost for the service and will you have to prepay six to twelve months in advance? (Hopefully the central station will be in business that long.)

**HOW TO PURCHASE AN ALARM SYSTEM
For Your Home**

Installing an alarm system can very expensive. You must be prepared to ask the right questions to get the most for your money.

The worst time to buy an alarm system is immediately after your home has been burglarized. Your are at the mercy of the burglar alarm salesman. They know you will be buying on emotion rather than on reason. An unscrupulous salesman could sell you equipment or services you might not need. And you could pay more than necessary for your alarm system.

Professionally Installed Alarm Systems

Be suspicious of claims like: "complete security", "absolute safety" or "this system is 100% effective in stopping an intruder". Some security systems are better than others but none can make these claims.

ALARM SYSTEM COSTS

Besides the initial costs of installing an alarm system, there are periodic costs and expenses as well. Here is a list of both:

Initial Costs

1) **Home survey and planning costs** - Many times these are not shown separately but are included in the total system price.

2) **Equipment costs** - Switches, sensors, master control panel, access panels, transmitters, receivers, automatic telephone dialer, digital communicator, local alarm bells and sirens.

3) **Installation costs** - Labor to mount and connect the equipment and components described above. Also includes installation supplies.

4) **Installation of special telephone lines and jacks** for the automatic telephone dialer or digital communicator.

5) Building permits and alarm system license fees.

Periodic Costs

1) Telephone company service charges.

2) Central station monitoring charges.

Home Security Systems

3) Response charges for a private security patrol.

4) False alarm fines imposed by some cities and police departments.

5) Service and repair costs.

PAYING FOR THE ALARM SYSTEM

If you cannot afford to pay cash for your alarm system you can probably finance it. Some alarm companies may offer to finance the system themselves or help find a lender who will.

Instead of financing it through the alarm company, try borrowing the money at your local bank, credit union or savings and loan association. These lenders might lend you the money with little or no collateral, but may require a lien on your house to insure payment.

As an alternative to borrowing, you might be able to lease the alarm system. Some alarm companies will lease the system directly or arrange to have a financial institution buy the system and then lease it back to you. A deposit may be required, but the payments can be kept within your budget. Leasing is another method of financing your alarm system making it more affordable.

If you decide to lease, make sure you understand all the provisions in the contract. Find out who is responsible for maintenance and repairs; you or the lessor. In addition, determine if you can purchase the alarm system at the end of the lease and what the purchase price will be.

Ask these questions about financing the alarm system:

1) Will the alarm company finance the system or can they recommend a lender who will?

2) Can you obtain your own financing from your bank, credit union or savings and loan association?

3) Can you arrange an unsecured loan or will a lien be placed on your home?

4) What recourse will you have against the alarm company if the alarm system does not work properly and you must continue to make payments?

5) Can you lease the alarm system with an option to buy?

After the installation is complete, obtain a release from the alarm company and the installers. This release can prevent a person or supplier who has furnished labor or materials for the alarm system from latter claiming he was not paid and placing a lien on your home.

FINAL WALK-THROUGH

When the installers have finished and before you hand over the final payment, get a complete walk- through of your new alarm system. Have them show you exactly how every component works and how to determine when one is not functioning properly. Also, get a complete set of written instructions so you will have a reference after the installers are gone.

Finally, make sure every member of your family is familiar with the alarm system and understands how to operate it.

Summary

It is unfortunate that we live in a time where we have to take those extra security precautions to protect our families, homes and possessions. If there is any one thing you have learned from this book I hope you have realized that the best time to plan and install an alarm system is before you need it.

While no home is completely burglar-proof and no car is completely safe from thieves, the advantages of installing an alarm system in your home and auto are far greater than doing nothing. Make a criminal's job as hard as possible and he will probably leave your home and auto alone.

Security experts estimate that only 5% of the homes in America are equipped with a security alarm system but the number is growing. Also, there are indications that vehicle alarm systems will be standard equipment in the near future. I hope you won't be left behind. Take the time to evaluate the real possibility of having your home broken into or car stolen.

WARNING - DISCLAIMER

This book is intended to be an introduction to home and vehicle alarms systems. Before you install an alarm in your home or vehicle, check with your city, county and state for laws and regulations governing their installation and use. Always follow the building and electrical codes and obtain the proper licenses and permits. If you have any problems selecting or installing alarm components, talk to a security consultant or alarm dealer. Also, hire a professional installer and licensed electrician if necessary.

The illustrations, diagrams and examples in this book should only be used as a guide and may not be applicable to your individual security requirements. The author and publisher have neither liability nor responsibility to any person, persons, or entity with respect to any loss or damage caused or alleged to be caused, directly or indirectly, by the information contained in this book.

In addition, no particular product, company, association or brand is recommended. The book is intended to be a introduction to home and vehicle alarm systems. Thoroughly study and check out any alarm company, security consultant, entity or individual you intend to deal with when purchasing and installing a security system.

I hope this book has actually helped you install an alarm system whether you did it yourself or hired a professional.

Terms & Definitions

Access Mode - The operation of an Alarm System such that no Alarm Signal is given when the Protected Area is entered; however, a signal may be given if the Sensors, Annunciator or Control Unit is tampered with or opened.

Active Sensor - A sensor that detects the disturbance of a radiation field that is also generated by the sensor. See also Passive Sensor.

Air Gap - The distance between two magnetic elements in a magnetic or electromagnetic circuit, such as between the core and the armature of a relay.

Alarm - An Alarm Device or an Alarm Signal

Alarm Circuit - An electrical circuit of an alarm system which produces or transmits an Alarm Signal.

Alarm Condition - A threatening condition, such as an intrusion, fire or holdup sensed by a Detector.

Alarm Device - A device that signals a warning in response to an Alarm Condition, such as a bell, siren or Annunciator.

Alarm Discrimination - The ability of an alarm system to distinguish between those stimuli caused by an Intrusion and those which are part of the environment.

Alarm Signal - A signal produced by a Control Unit indicating the existence of an Alarm Condition.

Alarm State - The condition of a Detector that causes a Control Unit in the Secure Mode to transmit an Alarm Signal.

Alarm System - An assembly of equipment and devices designed and arranged to signal the presence of an Alarm Condition requiring urgent attention such as unauthorized entry, fire, temperature rise, etc..

Annunciator - An alarm monitoring device that consists of a number of visible signals such as "flags" or lamps, indicating the status of the detectors in an alarm system or systems. Each circuit in the device is usually labelled to identify the location or condition being monitored. In addition to the visible signal, an audible signal is usually associated with the device. When an alarm condition is reported, a signal is indicated visibly, audibly or both. The visible signal is generally maintained until reset, either manually or automatically.

Area Protection - Protection of the inner space or volume of a secured area by means of a Volumetric Sensor.

Area Sensor - A sensor with a detection zone which approximates an area, such as a wall surface.

Audible Alarm Device - (1) A noise making device such as a siren, bell or horn used as part of a local alarm system to indicate an Alarm Condition. (2) A bell, buzzer, horn or other noise making device used as a part of an Annunciator to indicate a change in the status of an operating mode of an alarm system.

Audio Monitor - An arrangement of amplifiers and speakers designed to monitor the sounds transmitted by microphones located in the Protected Area. Similar to an Annunciator, except that supervisory personnel can monitor the protected area to interpret the sound.

Authorized Access Switch - A device used to make an alarm system or some portion or zone of a system inoperative in order to permit authorized access. A shunt switch is an example of such a device.

Burglary - The unlawful entering of a structure with the intent to commit a felony or theft therein.

Carrier Current Transmitter - A device that transmits Alarm Signals from a sensor to a Control Unit via the standard ac power lines.

Circumvention - The defeat of an alarm system by the avoidance of its detection devices. This is done by jumping over a pressure sensitive mat, by entering through a hole cut in an unprotected wall rather than through a protected door, or by keeping out of range of a detector or sensor. Circumvention contrasts with Spoofing.

Closed Circuit System - A system in which the sensors of each zone are connected in series so that the same current exists in each sensor. When a activated sensor breaks the circuit or the connecting wire is cut, an alarm is transmitted for that zone.

Contact - (1) Each of the pair of metallic parts of a switch or relay that by touching or separating make or break the electrical current path. (2) A switch-type sensor.

Control Unit - A device, usually electronic, that provides the interface between the alarm system and the human operator and produces an Alarm Signal when its programmed response indicates an Alarm Condition. Some or all of the following may be provided for: power for the sensor, sensitivity adjustments, means to select and indicate Access Mode or Secure Mode, monitoring for Line Supervision and Tamper Devices, timing of an alarm signal, etc.. (Same as Master Control Panel.)

Defeat - The frustration, counteraction or thwarting of an Alarm Device so that it fails to signal an alarm when a protected area is entered. Defeating includes both Circumvention and Spoofing.

Detection Range - The greatest distance at that a sensor will consistently detect an intruder under a standard set of conditions.

Detector - A sensor such as those used to detect Intrusion, equipment malfunctions or failure, rate of temperature rise, smoke or fire.

Dialer - See Telephone Dialer, Automatic.

Duress Alarm Device - A device that produces either a Silent Alarm or Local Alarm under a condition of personal stress such as a holdup, fire, illness, or other panic or emergency. The device is manually operated and may be fixed or portable.

Entrance Delay - The time between activating a sensor on an entrance door or gate and the sounding of a Local Alarm or transmission of an Alarm Signal by the Control Unit. This delay is used if the Authorized Access Switch is located within the Protected Area and permits a person with the control key to enter without causing an alarm. The delay is provided by a timer within the Control Unit or Panel.

Exit Delay - The time between turning on a control unit and the sounding of a Local alarm or transmission of an Alarm Signal upon activation of a sensor on an exit door. This delay is used if the Authorized Access Switch is located within the protected Area. This permits a person with the control key to turn on the alarm system and to leave through a protected door or gate without causing an alarm. The delay is provided by a timer within the Control Unit.

False Alarm - An alarm signal transmitted in the absence of an Alarm Condition. These may be classified according to causes: e.g., rain, fog, wind, hail, lightning, temperature, etc.; animals, e.g. rats, dogs, cats, insects, etc.; man-made disturbances, e.g., sonic booms, electromagnetic interference, vehicles, etc.; equipment malfunctions, e.g., transmission errors, component failure, etc.; operator error; and unknown.

Foil - Thin metallic strips which are cemented to a protected surface (usually glass in a door or window), and connected to a closed electrical circuit. If the protected material is broken so as to break the foil, the circuit opens, initiating an alarm signal. Also called tape. A window, door or other surface to which foil has been applied is said to be taped or foiled.

Heat Sensor - (1) A sensor that responds to either a local temperature above a selected value, a local temperature above increase, which is at a rate of increase greater than a pre-selected rate (rate-of-rise), or both. (2) A sensor which responds to infrared radiation from a remote source such as a person.

Infrared (IR) Motion Detector - A sensor which detects changes in the infrared light radiation from parts of the Protected Area. Presence of an intruder in the area changes the infrared light intensity from his direction.

Interior Perimeter Protection - A line of protection along the interior boundary of a Protected Area including all points through which entry can be effected.

Intrusion - Unauthorized entry into the property of another.

Intrusion Alarm System - An alarm system for signaling the entry or attempted entry of a person or an object into the area or volume protected by the system.

Ionization Smoke Detector - A Smoke Detector in which a small amount of radioactive material ionizes the air in the sensing chamber, thus rendering it conductive and permitting a current to flow through the air between two charged electrodes. This effectively gives the sensing chamber an electrical conductance. When smoke particles enter the ionization area, they decrease the conductance of the air by attaching themselves to the ions causing a reduction in mobility. When the conductance is less than a predetermined level, the detector circuit responds.

Jack - An electrical connector which is used for frequent connect and disconnect operations; for example, to connect an alarm circuit at an overhang door.

Local Alarm - An alarm that when activated makes a loud noise (see Audible Alarm Device) at or near the Protected Area or floods the site with light or both.

Loop - An electrical circuit consisting of several elements, usually switches connected in series.

Line Supervision - Electronic protection of an Alarm Line accomplished by sending a continuous or coded signal through a circuit. A change in the circuit characteristics, such as a change in impedance due to the circuit having been tampered with, will be detected by the monitor. The monitor initiates an alarm if the change exceeds a predetermined amount.

Magnetic Switch - A switch that consists of two separate units: a magnetically actuated switch and a magnet. The switch is usually mounted in a fixed position (door jamb or window frame) opposing the magnet, which is fastened to a hinged or sliding door, window, etc.. When the movable section is opened, the magnet moves with it actuating the switch.

Master Control Panel - See Control Unit.

Mat Switch - A flat area switch used on open floors or under carpeting. It may be sensitive over an area of a few square feet or several yards.

Mechanical Switch - A switch in which the Contacts are opened and closed by means of a depressible plunger or button.

Microwave Alarm System - An alarm system that employs Radio Frequency Motion Detectors, operating in the Microwave Frequency region of the electromagnetic spectrum.

Microwave Frequency - Radio frequencies in the range of approximately 1.0 to 300 GHz.

Monitoring Station - The Central Station or other area that guards, police or commercial service personnel observe Annunciators and registers reporting on the condition of alarm systems.

Motion Sensor - A sensor that responds to the motion of an intruder.

NICAD - (Contraction of "nickel cadmium") A high performance, long-lasting rechargeable battery (with electrodes made of nickel and cadmium), that may be used as an emergency power supply for an alarm system.

Normally Closed (N.C.) Switch - A switch in that the Contacts are closed when no external forces act upon the switch.

Normally Open (N.O.) Switch - A switch where the Contacts are open (separated) when no external forces act upon the switch.

Open Circuit System - A system where the sensors are connected in parallel. When a sensor is activated, the circuit is closed, permitting a current which activates an Alarm Signal.

Panic Button - (Also Panic Switch) See Duress Alarm Device.

Passive Sensor - A sensor that detects natural radiation or radiation disturbances, but does not itself emit the radiation on which its operation depends.

Perimeter Protection - Protection of access to the outer limits of a Protected Area, by means of physical barriers, sensor on physical barriers, or exterior sensors not associated with a physical barrier.

Photoelectric Beam Type Smoke Detector - A Smoke Detector that has a light source that projects a light beam across the area to be protected onto a photoelectric cell. Smoke between the light source and the receiving cell reduces the light reaching the cell, causing actuation.

Photoelectric Sensor - A device that detects a visible or invisible beam of light and responds to its complete or near complete interruption.

Protected Area - An area monitored by an alarm system or guards, or enclosed by a suitable barrier.

Radio Frequency Motion Detector - A sensor that detects the motion of an intruder through the use of a radiated radio frequency electromagnetic field. The device operates by sensing a disturbance in the generated RF field caused by an intruder's motion, typically a modulation of the field referred to as a Doppler Effect, which is used to initiate an Alarm Signal. Most radio frequency motion detectors are certified by the FCC for operation as "field disturbance sensors". Units operating in the Microwave Frequency range are usually called Microwave Motion Detectors.

Reed Switches - A type of Magnetic Switch consisting of contacts formed by two thin movable magnetically actuated metal vanes or reeds, held in a normally open position within a sealed glass envelope.

Remote Alarm - An Alarm Signal that is transmitted to a remote Monitoring Station. See also Local Alarm.

Reset - To restore a device to its original (normal) condition after an alarm or trouble signal.

Robbery - The felonious or forcible taking of property by violence, threat or other overt felonious act in the presence of the victim.

Secure Mode - The condition of an alarm system in which all sensors and Control Units are ready to respond to an intrusion.

Sensor - A device that is designed to produce a signal or other indication in response to an event or stimulus within its detection zone.

Shunt - (1) A deliberate shorting-out of a portion of an electric circuit. (2) A key operated switch that removes some portion of an alarm system for operation, allowing entry into a Protected Area without initiating an Alarm Signal. A type of access switch.

Silent Alarm - A Remote Alarm without a obvious local indication that an alarm has been transmitted.

Smoke Detector - A device that detects visible or invisible products of combustion. See also Ionization Smoke Detector and Photoelectric Beam Type Smoke Detector.

Spoofing - The defeat or compromise of an alarm system by "tricking" or "fooling" its detection devices by short circuiting part or all of a series circuit, cutting wires in a parallel circuit, reducing the sensitivity of a sensor, or by entering false signals into the system. Spoofing contrasts with Circumvention.

Standby Power Supply - Equipment that supplies power to a system in the event the primary power is lost. It may consist of batteries, charging circuits, auxiliary motor generators or a combination of these devices.

Supervised Lines - Interconnecting lines in an alarm system which are electrically supervised against tampering. See also Line Supervision.

Supervisory Circuit - An electrical circuit or radio path that sends information on the status of a sensor or guard patrol to an Annunciator. For Intrusion Alarm Systems, this circuit provides Line Supervision and monitors Tamper Devices.

Tamper Device - (1) Any device, usually a switch, that is used to detect an attempt to gain access to the intrusion alarm circuitry, by removing a switch cover. (2) A monitor circuit to detect any attempt to modify the alarm circuitry, such as by cutting a wire.

Tamper Switch - A switch that is installed in such a way as to detect attempts to remove the enclosure of some alarm system components such as: control box doors, switch covers, junction box covers or bell housings. The alarm component is then often described as being "tampered".

Telephone Dialer, Automatic - A device that, when activated, automatically dials one or more pre-programmed telephone numbers, (e.g., police, fire department) and relays a recorded voice or coded message giving the location and nature of the alarm.

U.L. Listed - Signifies that production samples of the product have been found to comply with established Underwriters Laboratories requirements. The manufacturer is authorized to use the Laboratories' listing marks on the listed products which comply with the requirements, contingent upon the follow-up services as a check on compliance.

Ultrasonic - Pertaining to a sound wave having a frequency above that of audible sound (approximately 20,000 Hz). Ultrasonic sound is used in ultrasonic detection systems.

Volumetric Sensor - A sensor with a detection zone which extends over a volume such as an entire room, or part of a room or passage way. Ultrasonic Motion Detectors are an example of volumetric sensors.

Zoned Circuits - A circuit that provides continual protection for parts or zones of the Protected Area while normally used doors and windows or zones may be released for access.

Zones - Smaller subdivisions into which larger areas are divided to permit selective access to some zones, while maintaining other zones secure and to permit pinpointing the specific locations from which an Alarm Signal is transmitted.

Index

Access panels
  81,83-85,121
  installation 133
  purchasing 129

Alarm components
  Auto 21,38,61
  Home 126-129

Alarm messages/signals
  20,72,73,74,108,112

Alarm ordinances 48

Alarms/alarm devices
  Home 72
  Auto 20,27,38,43,106-108

Alarm status lights 81

Annunciators 81

Area protection
  67,68-69,115,118-120

Area sensors/detectors
  94-104
  installation 145-149

Audio sensors/detectors
  see sound discriminators

Automatic reset 81

Automatic shutoff 81

Automatic telephone dialer
  72,74,80,107-108,111-112
  locating 120-121
  purchasing 127-128

Batteries 45,83

Battery backup systems
  Auto 45
  Home 83,107

Bells
  Auto 20,29
  Home 72,107
  also see local alarms

Carrier current 73
  also see line carriers

Central station 105,108,112-113

Central station monitor
  78,108,112-113

Central station monitoring
  112-113
  questions to ask 167

Circuits
  day/night 82
  delay entry/exit 21,43,81,85
  fire protection 75,82,105
  normally closed (N.C.) 80,87
  normally open (N.O.) 78,87
  reset 23
  supervised 78,82
  24 hour 82

Combination panels 77

Control unit/panel
  18,20,21,38,39,40,43,60

Digital communicators
  72,74,80,108,120-121
  purchasing 127-128

Dead bolts (for auto) 40,51

185

Digital key pad
    Auto 32,38,44
    Home 85

Fail safe arming 83

False alarms
    Auto 47-48
    Home 89,94,112,120,158

Floor mat switches
    101,102,104
    installation 147

Floor plans 114-115,125

Fuel cutoff 50

Fuel pump cutoff 50

Fuel pump lockout 24

Glass breakage sensors
    89,92
    installation 139-140

Hard-wire systems 152,156

Hard-wiring 72,134,152,154

Heat detectors
    69,82,104,105-106,120
    installation 151

Hood locks 51

Ignition cutoff 23,49-50,60

Installation supplies
    Auto 38
    Home 130-131

Installation tools
    Auto 38
    Home 130

Latching relays 21

Line carriers 72,73,145,151

Local alarms
    74,106,108,111,121
    installation 133-134
    purchasing 127-128

Lockup relays 21,80

Loops see circuits

Magnetic switches 85-87,92
    installation 134-138

Master control panel
    70,72,75-83
    installation 131-133
    locating 120-121
    purchasing 126-127

Medical switches 70

Microwave detectors/
    sensors 77,94,96-97
    installation 146,149

Motion detectors 20,25,43,47

National Burglar and Fire
    Alarm Association
    (NBFAA) 162

Override switch 23

Paging systems
    20,21,30,43-44,60,61

Panic switches
  Auto 27,35
  Home 70,104,120,149-150

Passive infrared detectors/
  sensors 98,99
  installation 146,149-150

Perimeter protection
  67,68,115,116-117

Perimeter switches/sensors
  installation 134

Permits/licenses 123,161

Photoelectric beam
  detectors/sensors
  98,100,101
  installation 147,150

Pin switches 20,25,43

Plunger switches 87-89
  installation 138

Point of entry protection
  see perimeter protection

Pressure sensitive mats
  see floor mat switches

Reed switches 87
  also see magnetic switches

Reset switches 23,80

Shock sensors
  Auto 25
  Home 89,92,139-140

Shunt switches 85

Sirens
  Auto 27,40,38
  Home 72,74,106
  also see local alarms

Smoke detectors
  69,82,104,105,120
  installation 150,151

Sound discriminators
  Auto 25,43
  Home 68,77,94,95,145

Space protection
  see area protection

Special protection switches/
  sensors 104-106

Special protection systems
  67,69-70

Standby batteries
  see battery backup systems

Starter lockout/cutoff 24,50

Status lights 85

Take-off connectors 143

Tamper switches 85,106

Testing
  Auto 47
  Home 157

Ultrasonic detectors/sensors
  Auto 27
  Home 77,94,97-98,146-148

Underwriters Laboratories
(UL) 125,165-166

Vehicle immobilizers
see ignition cutoff
see electronic fuel pump cutoff
see fuel cutoff
see fuel pump lockout

Voice synthesizers 29

Voltage drop sensors 20,25,43

Walk test light 146

Warning signs 55,159

Window foil 89,91-92
installation 139,141,143

Window glass etching 53

Wired window screens 92-93
installation 143-145

Wireless transmitter systems
152,153

Wireless transmitters &
receivers 72-73
purchasing 129

Wiring
Auto 45-46
Home 121-122,151

Zones 77,78

Colophon

This book is an example of desktop publishing. Except for the illustrations, the entire book was composed, edited and typeset on a microcomputer.

Production Notes:

Drafting & Editing:
Computer - IBM Personal Computer
Software - Microsoft Word

Typesetting:
Printer - Hewlett Packard LaserJet
Typestyle - Helvetica (titles), Times Roman (text)

Paper:
Text - 60# white offset book
Cover - 10 pt.

Ink:
Text - Standard black
Cover - 2 color with Delta Diamond Coat

Printing:
Offset by Delta Lithograph , Van Nuys, California

Binding:
Perfect (adhesive)

Why Plastic Windows Are Better Than Glass

By Doug Kirkpatrick

The key to protecting any home against burglars and intruders starts by securing every door and window. As you have seen, home security systems do this with perimeter sensors and detectors. But because it is so easy for an intruder to break a window and climb inside, many homeowners want additional protection to suppliment their alarm systems.

My new book titled "**How to Install Burglarproof - Energy Efficient Windows**" shows you how to install window panes that are virtually unbreakable. Unlike iron burglar bars, windows made from space-age plastic won't make your home look or feel like a jail cell. And plastic window panes won't trap you inside your home if a fire were to break out.

Besides deterring break-ins and preventing accidental breakage, windows made from high-tech plastic insulate a home better than glass. These benefits protect your home and family while saving you money on your utility bills.

To order your copy of this book, just use the order form on the next page.

Order Form

Please send me the following books by Doug Kirkpatrick:

_____ Copies of **The Complete Guide to Home & Auto Burglar Alarms** @ $12.95 each (U.S. Funds)

_____ Copies of **How to Install Burglarproof - Energy Efficient Windows** @ $4.95 each (U.S Funds)

I understand that I may return any book for a full refund if I am not completely satisfied.

Name:_____

Address:_____

City:_____State:_____Zip:_____

Californians: Please add 6 1/2% sales tax.

Shipping: Please add $1 for the first book and .50 for each additional book.

_____ I can't wait for surface shipping. Here is $2.50 per book for air mail. (For Complete Guide to Home & Auto Burglar Alarms only.)

Send to:

Baker Publishing
Post Office Box 8322-K
Van Nuys, CA 91409

(Just photocopy this page or write your order on a piece of paper)

Order Form

Please send me the following books by Doug Kirkpatrick:

_____ Copies of **The Complete Guide to Home & Auto Burglar Alarms** @ $12.95 each (U.S. Funds)

_____ Copies of **How to Install Burglarproof - Energy Efficient Windows** @ $4.95 each (U.S. Funds)

I understand that I may return any book for a full refund if I am not completely satisfied.

Name:_____

Address:_____

City:_____State:_____Zip:_____

Californians: Please add 6 1/2% sales tax.

Shipping: Please add $1 for the first book and .50 for each additional book.

_____ I can't wait for surface shipping. Here is $2.50 per book for Air Mail. (For Complete Guide to Home & Auto Burglar Alarms only.)

Send to:

Baker Publishing
Post Office Box 8322-K
Van Nuys, CA 91409

(Just photocopy this page or write your order on a piece of paper)